Preface

This collection of articles reflects the diversity of interests and personalities of the membership of the British Pteridological Society in its hundredth year and recounts its history to date. Above all, this volume shows that ferns, and the rather cumbersomely named fern allies, have attracted and will continue to absorb the attention of people - whether at the molecular, taxonomic, horticultural or metaphysical level - over more than the century we celebrate in 1991. The present membership has no doubt that our successors will be celebrating their enjoyment of ferns and still be assessing the past, present and future of these amazing plants on many occasions. We raise our glasses to pteridophytes and you!

I am pleased to report that all contributors volunteered that they *enjoyed* writing these articles. I thank them all for their willing response. I have added biographical notes on the authors, based on information they provided, to all the articles except the three that contain an autobiographical element.

The decorative line illustrations are taken from W. & J. Birkenhead (1890) *Catalogue of Ferns* and J. Birkenhead (1892) *Ferns and Fern Allies*.

My thanks go also to Jim Crabbe, Clive Jermy and Barry Thomas for their help and efforts to teach me the rudiments of editing, to Paul Lund for producing the photographic prints, and to all those, but especially Alison Paul, Jim Crabbe and Nick Hards who helped with proof-reading.

Josephine M. Camus

100 Years

Barry A. Thomas
President

Ferns have always been popular plants in the home and garden. They are usually thought of as plants with finely divided leaves that grow in shady places, but in reality they are much more diverse in shape and in the places they grow. In fact, there are many beautiful and spectacular ferns to be found in countries throughout the world and a great number of these have been brought into cultivation for the enthusiast.

The real craze for growing ferns started in the middle of the 19th century and very quickly swept throughout Victorian Britain. Everyone seemed to want ferns and there was much to encourage this. It was a time of active plant hunting that was bringing a constant stream of new species and varieties into cultivation.

The British Pteridological Society was founded in 1891, in the English Lake District, by a small group of fern enthusiasts who wished to channel their common interest through meetings and publications. They also wanted to ensure that the accumulated knowledge and the wealth of fine varieties in cultivation were not lost to future generations.

The scientific study of ferns was also well under way by this time, although there were still many important discoveries waiting to be made. Many of these revelations were to appear later in the Society's publications.

One hundred years later the Society is still flourishing and has broadened to include both amateur and professional botanists and gardeners from all over the world. It caters for all those interested in the taxonomy, classification, evolution, ecology and conservation of ferns as well as for the traditional enthusiasm for growing them. Its regular and special publications contain a wealth of information for all who are interested in ferns, clubmosses or horsetails. The members organise national and regional field meetings in Britain, talks and discussions, garden visits and plant and spore exchange schemes. The Society also supports amateurs undertaking research on pteridophytes.

This centenary year is clearly a special one for the Society. To mark it we have produced this anthology of articles and illustrations to chart our first hundred years. In it you will find details of both the Society and its members and the changes in scientific and horticultural knowledge of ferns that have occurred during this time.

We will leave future members to take care of the Society and pursue its aims of encouraging a wide interest in ferns. Ferns are here to stay and I trust that the Society will last with them.

The History of British Pteridology 1891-1991

CENTENARY
1891~1991

The British Pteridological Society

London

1991

Published by:
The British Pteridological Society
c/o The Natural History Museum
Cromwell Road
London SW7 5BD
England

Special Publication No. 4
Edited by J. M. Camus

Cover design by Richard Rush
ISBN 0 9509806 3 3
© British Pteridological Society
1991

The BRITISH PTERIDOLOGICAL SOCIETY was founded in 1891 and today continues as a focus for fern enthusiasts. It provides a wide range of information about ferns through the medium of its publications and available literature. It also organises formal talks, informal discussions, field meetings, garden visits, plant and spore exchange schemes and fern book sales. The Society has a wide membership which includes gardeners, nurserymen and botanists, both amateur and professional. The Society's journals the *Fern Gazette, Pteridologist* and *Bulletin* are published annually. The *Fern Gazette* publishes matter chiefly of specialist interest on international pteridology, the *Pteridologist* topics of more general appeal, and the *Bulletin* Society business and meetings reports.

Contents

Prologue
Salute to Ferns — 6
B. W. Graham

Pteridophytes and People
The Study of Fossil Ferns — 7 - 15
B. A. Thomas

Changes in the List of British Pteridophytes — 17 - 24
J. M. Camus

Pteridophytes as Indicators of Landscape Changes in the British Isles in the Last Hundred Years — 25 - 40
C. N. Page & H. S. McHaffie

My Life with Ferns — 41 - 43
R. E. Holttum

An Amateur's Steps in Pteridology — 45 - 48
C. R. Fraser-Jenkins

A Brief History of Ferns and their Cultivation — 49 - 57
M. V. Ford

The Development of Laboratory Based Studies in Fern Variation — 59 - 63
M. Gibby

The Story of the Reginald Kaye Fern Collection — 65 - 69
R. Kaye

Conservation – the Fern Story — 71 - 74
M. H. Rickard

One Hundred Years of Illustrated Fern Books (and then some); a personal review — 75 - 81
P. G. Barnes

The Society
The British Pteridological Society – the First Hundred Years
Appendix of Officers 1891 - 1991 — 83 - 93
J. W. Dyce

Gleanings from the Minute Book 1891 - 1991 — 95 - 103
A. R. Busby

Plates — 104 - 118

The Presidents of the British Pteridological Society — 119 - 126
N. A. Hall

Epilogue
Testament of a Fern Lover — 127
R. J. Smith

Salute to Ferns

When the world was formed from the cosmic dust,
As it circled the nascent sun,
Below the layer of hardening crust
Evolution had scarce begun.

The millennia passed at an ordered pace,
Equipping the primal soup
To compete in the great survival race,
By division of group from group.

The critical hour came round in time
When the first blue-green appeared,
Then, emerging from beds of oozing slime,
The crook of a frond was reared.

Arcane carboniferous forests grew,
As reptiles crept out of the sea,
While the hungry *Archaeopteryx* flew
Among ferns the height of a tree.

The ferns saw the dinosaurs come and go,
And hominids take their place.
They suffered in heat and drought and snow,
But they kept well up in the race.

The civilisations came and went,
Aggression was fierce on all sides,
Though unarmed, and unsure of human intent,
It wasn't the ferns that died.

Despite being plundered to meet the demand
Of the late Victorian craze,
The charm of pteridophytes must command
Our respect, admiration and praise.

How right it is we should humbly observe
That the fittest indeed survive.
If species become what they richly deserve,
The ferns will continue to thrive.

On the shape of the future we speculate,
As the shroud of Nemesis nears,
But today we are gathered to celebrate
The coming and past hundred years.

Bridget Graham
Polpey, Par, Cornwall PL24 2TW

Bridget Graham's interests are very diverse. Trained as a pharmaceutical dispenser, she supports the mentally handicapped, adores ballet, works tapestry, and has a keen interest in the theories proposed to explain the nature of the Universe, as well as being an accredited poetess. She found time during the second World War to become interested in plants and has been a fern enthusiast since the mid-1960s.

The Study of Fossil Ferns

Barry A. Thomas
Department of Botany
National Museum of Wales, Cardiff CF1 3NP

According to the Code of Botanical Nomenclature (Greuter et al., 1988), palaeobotanical literature commenced in 1820, so only 71 years of published research preceeded the birth of our Society. In 1891, the year of our foundation, the eminent British palaeobotanist, Dr Robert Kidston, published a timely account that summarised knowledge of the older British Palaeozoic species of fern. Unfortunately, many of his interpretations have proved to be incorrect, simply because the general assumption at that time that everything of fern-like appearance was a fern is wrong. Many of the Palaeozoic plants with fern-like foliage were subsequently shown to belong to a group intermediate between the ferns and the gymnosperms. Although it had been suspected for some time that such an intermediate group of plants had once existed, the first proof of the plants came from the work of Oliver and Scott (1904) when they described the seed-ferns or pteridosperms. These included such common Carboniferous fern-like genera as *Neuropteris* and *Mariopteris*. Much later other fern-like foliage was shown to be different to that of both ferns and pteridosperms. In 1962, Beck proved the anatomy of the fern-like *Archaeopteris* to be gymnospermous, even though the free-sporing habit of the fertile foliage was in essence pteridophytic. These plants, the progymnosperms, are now known to have been both common and diverse in the Palaeozoic and are believed to be the progenitors of the true gymnosperms.

Soon after Kidston published his review in 1891, Sir Albert Seward (1894, 1900) also published two important works; this time on British Mesozoic plant fossils. These two authors gave much of the factual information that was available at the time and their work, together with that of others, such as Nathorst (1906) and Thomas (1911), was later used by Bower (1923-28) in his monographic work on ferns. Bower usefully charted the evolution of the families of the more primitive ferns, such as the Marattiaceae, Osmundaceae, Schizaeaceae, Matoniaceae and Gleicheniaceae but, as Harris (1973) has pointed out, there was little evidence of plant fossils available to guide Bower in his interpretation of the more advanced polypodiaceous ferns. More recent palaeobotanical studies have also shown some of the earliest filicalean ferns to have been incorrectly assigned to families of living ferns. Holttum (1982) has even been quite critical of much of Bower's work on living ferns regarding his generalisations (summarised in the 1928 volume) to be untenable, being based on too few species, too great an emphasis on vascular anatomy and on outdated 19th century taxonomy.

There were very few major advances in palaeobotanical studies on ferns for many years after Bower's work, although many papers were published that increased our general knowledge of the subject. Some described individual species, while others gave lists or illustrated assemblages of plants, so it is perhaps as well to remind the reader that many species of fossil plants are defined on very few characters and that different writers have different views on the validity of species characters. Harris (1961) assessed the situation correctly by noting that imperfect knowledge tends to multiply the number of fern species, even though it appears to have the opposite effect on the number of gymnosperm species.

It is perhaps in the last thirty years that most new discoveries have been made and an increasing number of papers published on fossil ferns. This has not only given a greater knowledge of their anatomy, morphology and systematic relationships but has given us some insight into their geological history, evolutionary relationships and ecologies. The following sections give a brief outline of the more important aspects of fossil ferns, indicating wherever possible how our interpretations have changed over the years. I have been very selective as it is impossible to do otherwise in the space available.

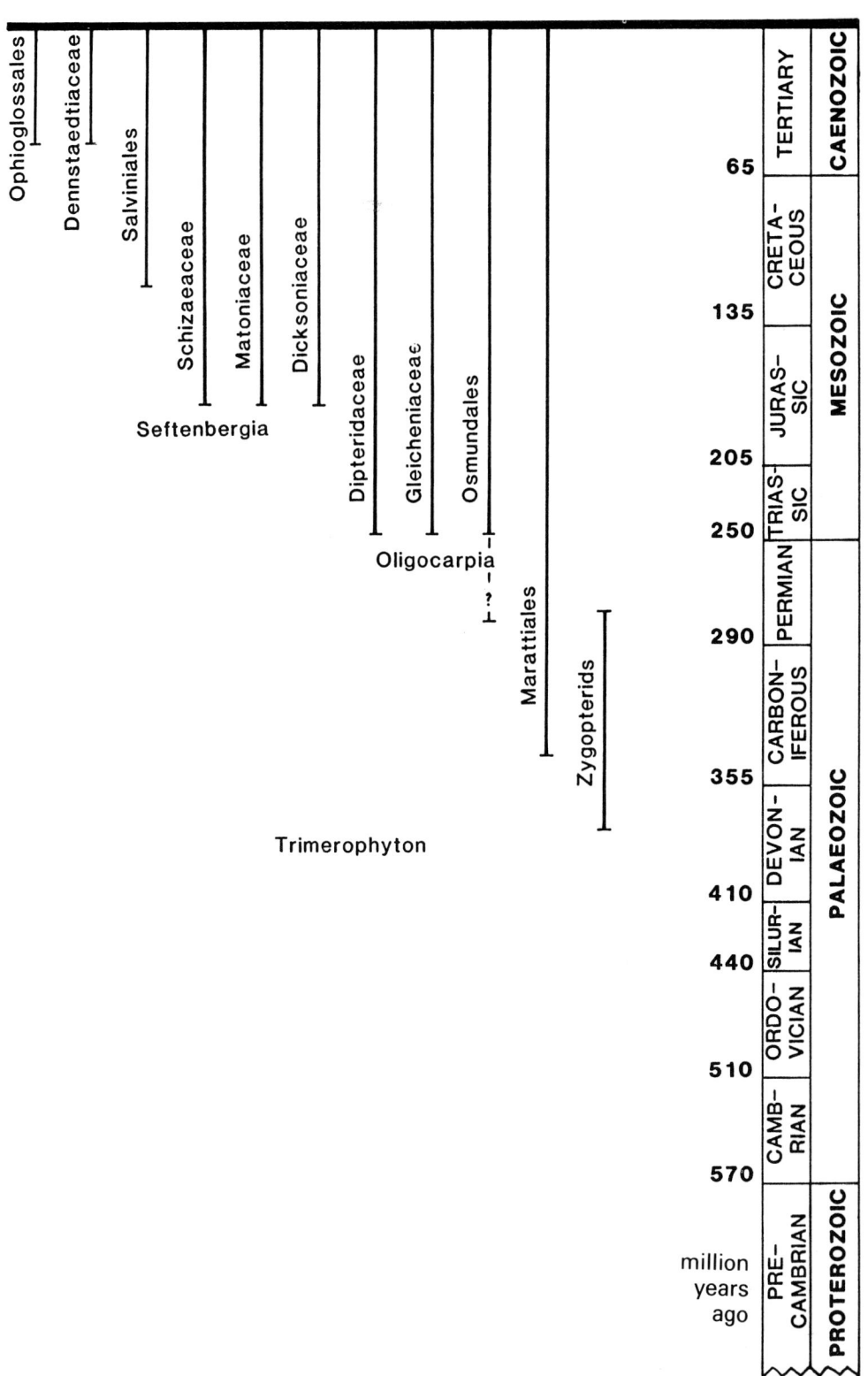

The fossil record of major fern groups

Techniques in studying fossil ferns

Several palaeobotanical techniques have enabled researchers to gain a much fuller understanding of plant fossils than was possible in the last century. The reader is referred to Lacey (1963) for the range of techniques available up to then. Some fossils are preserved as petrifactions having been infiltrated by mineral rich water before they decomposed. Although the early researchers were able to section such plant petrifactions and learn much about the internal anatomy of the plants preserved in this way, the wastefulness of the cutting and grinding processes meant that small organs were often represented by only one section. The study of such petrifactions was made much more practical with the development of the peel method, which allowed close serial sections to be prepared for the first time. This involves dissolving away with acid a few thousands of a millimetre of the matrix; a solution of cellulose acetate in acetone is poured on to dry over the protruding, acid-resistant cell walls of the fossil. The sheet, when dried, can be peeled off bringing the embedded fossil tissue with it (Walton, 1928). Even this technique has been improved by the rapid peel method of Joy, Willis and Lacey (1956), which uses cellulose acetate sheets rather than solutions as in Walton's original method.

The study of spores from sporangia is now an accepted part of palaeobotanical research and descriptions of spores appear in both generic and specific diagnoses. Yet it was only in the first decade of this century that acid maceration techniques were employed for preparing such spores.

When a fossil is formed through plant material being entombed in waterlogged silt and subsequently compacted by the increasing weight of overlying sediments it is called a compression. Often original plant material such as cutin is preserved. When such fossils are exposed through the splitting of the rock the fracture line will run over the smoother surface of the plant material. This results in the rougher surface, with such structures as sori and sporangia, being embedded in the rock and therefore largely invisible. However, in 1923 Walton developed his Canada balsam technique enabling a compression on a rock surface to be effectively 'turned over' allowing small stuctures such as sporangia to be seen. In this method, the fossil-bearing surface of the rock is attached to a glass slide with Canada balsam and the exposed glass coated with wax; the rock is then dissolved with hydrofluoric acid thereby exposing the entire fossil on the slide. Modern resins now make this process much simpler as the use of a glass slide (which would also be dissolved by the acid if not coated in wax) is not necessary. Light microscopes are now available to give better resolution via phase contrast, Nomarski differential interference contrast and fluorescence. There are also scanning electron microscopes to examine small specimens in great detail and transmission electron microscopes for ultrastructural studies of spores and cuticles. Biochemical analysis of plant fossils has not yet received all the attention it deserves and there is potentially much scope for this in the study of ferns (see Thomas, 1986, for a summary of methods and results). All of these techniques permit us to study plant fossils in greater detail, but it must be remembered that the discovery of new and suitable specimens still relies to a large extent on chance. Any reader unfamiliar with the study and interpretation of fossils, but wishing to know more, should read the accounts of fossilization and methods of study in general texts such as Stewart (1983), Taylor (1981), Thomas and Spicer (1987).

Early ferns

It is generally accepted that an early Devonian group, the Trimerophytes, gave rise to the ferns, cycads, conifers and ultimately the flowering plants. It was the monopodial branching patterns of these relatively simple plants that gave them such a potential for large scale evolutionary change and diversification. The earliest fern-like plants that arose from them are, however, poorly understood. The problem is often how to interpret small scraps of plants that show a small number of either vegetative or reproductive characters. Galtier and Scott (1985) have reviewed the diversification of the early ferns and suggest that many have features reminiscent of the Trimerophytes, because both Lower and Upper Carboniferous ferns are known to have very simple stem anatomy, unplanated leaves devoid of, or with reduced, laminate pinnules possessing terminal sporangia.

Some of the more specialised early ferns cannot be related to modern ferns. They evolved, diversified and disappeared without trace, for example the Zygopterids originated in the Devonian and disappeared in the Permian. Their extinctions may have been through unsuccessful competition with other plants or through climatic change, but we will probably never know the real cause for certain. Other early ferns possessed a more generalised type of morphology from which any type of modern fern could theoretically have evolved. Some of these possessed annulate sporangia very similar to those in living Osmundalean and Filicalean ferns.

Marattialeans

Although it has been assumed for many years that the Marattiales had their origin in the Palaeozoic, a large amount of recent work has confirmed that members of the Marattiaceae have existed from the Carboniferous onwards. Many papers have been published on the numerous Coal Measures genera (e.g. Stubblefield, 1984; Hill, Wagner & El-Khayal, 1985). The range of such plants from around the world shows that rapid evolutionary change was occurring within the group. Much discussion has centred on the idea that the Marattialean synangium could have evolved from an earlier branched, but lamina-less, frond with terminal sporangia. Bower (1926), Mamay (1950) and many others have referred to this evolutionary process as a phyletic slide. As more evidence becomes available from a greater variety of fossils our interpretations can become more detailed (e.g. Gao, 1988). By the Jurassic there were forms indistinguishable from modern genera, for example Harris (1961) included specimens from the Yorkshire Jurassic in *Marattia*. Much has been written about the evolutionary relationships of the various fossil taxa. Hill and Camus (1986) give a broad overview of the group and assess the evolutionary relationships of its member genera by cladistical methods.

Osmundales

It has been suggested that the group had its origins in the Upper Carboniferous and Permian genus *Grammopteris* (Miller, 1971) but the heavy reliance on sporangial characters makes the suggestion very open to debate. As in the Marattiales, there seems to have been a rapid evolutionary diversification in the Palaeozoic followed by more gradual changes later on. There are many Upper Permian genera of petrified stems, some with creeping rhizomes and others with erect stems covered with densely packed fronds. The first indisputable fertile foliage is known from the Triassic (Ash, 1969). This genus, *Todites*, is known from many later Mesozoic localities, but the gradual reduction in the number of taxa suggests that the group was already in decline. This seems to have continued to the present day for there are now only about 16 species in three genera.

Other Filicaleans

The filicalean type of annulate sporangium appears to have evolved by the Lower Carboniferous (approximately 355 million years ago). Rothwell (1987) has even described filicalean frond segments, from the Upper Carboniferous of eastern North America, that bore indusiate sori with gradate development of its sporangia. Based on this evidence, Rothwell suggests that the Filicales originated about the time of the boundary of the Devonian-Carboniferous periods and underwent evolutionary radiation resulting in diverse forms during the Carboniferous.

The timing of the first appearance of modern families has been a matter for much debate over the years. For example the Carboniferous *Senftenbergia* has sporangia with a prominent terminal annulus similar to those of extant Schizaeaceous ferns. Indeed, for many years, since Radforth (1938), it was generally accepted that the *Senftenbergia* form was the beginning of the Schizaeaceae *sensu lato*, leading to the Jurassic genera such as *Klukia* and forms very much like modern genera in the Cretaceous. Doubts about the reliance of using the terminal annulus as a diagnostic character have centred on the variability of its relative size and extent within the genus *Senftenbergia* as a whole. Furthermore in *Senftenbergia* the annulus is never just a single tier of cells as in most members of the Schizaeaceae, and other taxa of living ferns are known to have sporangia with a terminal annulus. Finally Jennings and Eggert (1977) solved the problem by describing a specimen of *Senftenbergia* with internal anatomy similar to that of the earlier ferns and quite unlike that of living members of the Schizaeaceae. Similarly the Carboniferous *Oligocarpia* is no longer thought to belong to the Gleicheniaceae. It now seems more likely that there were no species of Carboniferous ferns that resembled modern taxa closely enough to be included within them.

An evolutionary radiation between the Permian and the Jurassic did, however, lead to the establishment of several of the more primitive families of filicalean ferns. There were Triassic ferns that bear resemblance to the Schizaeaceae but, as they possessed reticulate venation and scattered sporangia, Ash (1969) preferred to create a new family for them. In contrast he included his genus, *Wingatea*, within the Gleicheniaceae and *Clathropteris walkeri* in the Dipteridaceae. Ferns rapidly diversified and increased in numbers throughout the Mesozoic and Harris (1961) thought they might have been the dominant herbs during the Jurassic. Other families were definitely present at this time as there are fossils referable to the Schizaeaceae *(Klukia)*, Matoniaceae (*Phlebopteris, Matonidium* and *Weichselia*), Dicksoniaceae (*Coniopteris*) and Dipteridaceae (*Clathropteris*).

Other fossil ferns are known from the Mesozoic that cannot be included in any families of living taxa. A very few of these have been thought to be polypodiaceous. The oldest, *Aspidistes*, had sori possessing small sporangia, each with an approximately vertical annulus, and a reniform indusium. The Middle Cretaceous *Gleicheniopsis* also had large numbers of small sporangia, this time shown to have 32 triradiate spores. However in reassessing these fossils, Harris (1973) did not consider their rather few characters sufficient to show the ferns to have undoubted polypodiaceous affinity. Possibly future work on late Mesozoic fossils may show some to have indisputable polypodiaceous characters.

Another evolutionary radiation started in the Cretaceous and appears to be continuing today (Lovis, 1977; Buckley, 1986). This has given rise to the variation of taxa within the families of polypodiaceous ferns. Even so there is still very little fossil evidence to give information on the geological history of these taxa. Records of Tertiary ferns are very scarce and most are of sterile foliage which cannot be referred to modern genera with any certainty. There are a few anatomically preserved remains of rhizomes and rachises that reveal the presence of the Dennstaedtiaceae in the Tertiary (Arnold & Daugherty, 1963; Ribbins & Collinson, 1978). Collinson (1978) has also described dispersed sporangial masses from the British Tertiary as belonging to the modern genus *Acrostichum*. The records are clearly very few, but it does not necessarily mean that ferns were scarce. The problem is more likely to be the result of the growth habit of the ferns, because most living polypodiaceous ferns have fronds that collapse, wilt and die on the plant. Therefore studies of dispersed fragments of ferns will probably prove to be one of the best ways of increasing our knowledge of polypodiaceous ferns in the Tertiary.

Ophioglossales

Although the modern taxa give the order virtual circumpolar distribution, until recently the only fossil record was of some early Jurassic and Cretaceous spores from the U.S.S.R. However, Rothwell and Stockey (1989) have recently described some specimens from the early Tertiary of Alberta, Canada that show sufficient characters to permit a definite assignment to *Botrychium*. In fact the overall morphology, pinnule shape and venation and arrangment of sporangia all show a very marked resemblance to those of *B. virginianum* which grows in Alberta today.

Salviniales

Very little was known of the fossil history of this group until relatively recently. Through the work of Jain and Hall (1969), Jain (1971) and Collinson (1980), we know that both *Salvinia* and *Azolla* have fossil records back to the Cretaceous. Most of the remains are of dispersed spores with the most distinctive being the megaspores of *Azolla*.

Reconstructions of whole plants

Only very rarely are whole plants found as fossils. Ash and Tidwell (1986) have described such a rare specimen from the Permian of New Mexico. This fern has a horizontal rhizome with roots, a short thick upright aerial branch and a tuft of dimorphic leaves. It is most probably a juvenile plant.

Most plants fragment before fossilization. The reconstruction of whole plants from a series of disarticulated organs or fragments is then only possible by finding the occasional organically connected pair of normally separated organs or through the discovery of specific anatomical similarities that can be taken as evidence of original organic union. Mere association is insufficient evidence.

Reconstructions of Carboniferous tree-ferns have been in the literature for well over a hundred years although many of these were based on very little hard evidence. *Psaronius* is a genus of Carboniferous tree-fern stems, although the name is also used for the whole plant. Many species have been described over the years although the early work concentrated mainly on species distinction. Recent work has tended to concentrate on the overall morphology of the plants and on their growth patterns. A restoration of a whole *Psaronius* has been illustrated by Morgan (1959) and a shoot apex with fronds by Stidd (1971). Other recent reconstructions of ferns include the Cretaceous tree-fern *Tempskya* (Andrews, 1948), the Cretaceous Matoniaceous *Weichselia reticulata* (Alvin, 1971), the Jurassic Osmundaceous *Todites princeps* (Schweitzer, 1978) and the Triassic Matoniaceous *Phlebopteris smithii* (Ash et al., 1982).

A study of the growth and development of extinct plants may also seem a natural progression from anatomical or morphological investigations but the lack of suitable specimens usually prevents any such interpretations being made. Some good examples are, however, available that demonstrate how valuable such work can be. For example, further serial sectioning of *Psaronius* has demonstrated an increasing complexity of vascular system from stem base to apex. From this, Mickle (1984) suggested that their apices continuously expanded forming new leaf and stem vascular bundles. There is also evidence that the bases of the stems sometimes decayed although growth continued as in some living ferns. Trivett and Rothwell (1988) stressed that such precise models of shoot development are becoming essential elements for reconstructing whole plants. By examining a Palaeozoic fern, *Anachoropteris clavata*, that has latent croziers replacing basal pinnae on mature frond segments, they suggested it grew as a liana. This interpretation was through a comparison with similar developmentally diagnostic features of the living climbing fern *Lygodium japonicum*.

Such studies as these are clearly valuable contributions to our understanding of whole plant biology. But, in addition, pictures of whole plant communities can be built up from such studies and from them an assessment of the changing patterns of plant community physiognomy.

Ecology

Early attempts at vegetational reconstruction were often simple groupings of plants based on associations of remains. Interpretations of the habitat relied upon knowledge of the morphological characters of some of the constituent plants. An appreciation of the relation of plant fossils to the sediments in which they are found has, however, led to a greater understanding of the relationships of the original communities to the sites of fossilization. This can be especially useful in the case of Carboniferous floras where there are many types of sedimentary rocks in relatively short sequences, e.g. Scott (1979).

Work on plant petrifactions (coal balls), found in large quantities in some North American coal seams, has resulted in similar findings. Here there is evidence that the Marattialean tree-ferns were widely spread in many habitats of the lowland Upper Carboniferous (Dimichele, Phillips & Peppers, 1985). The swamp communities there were dominated largely by arborescent lycophytes until the late Upper Carboniferous when they seem to have all but disappeared due to the drying out of the swamps. With the re-establishment of swampy conditions the tree-ferns together with the seed-ferns quickly colonised the areas, leaving the drier lowland ground dominated by the early conifers.

There are other examples of relatively recent work on the ecologies of ferns, but most are within interpretations of floras. The work on the common and widespread Lower Cretaceous *Weichselia* by Alvin (1974) is rather different because it deals with charred leaf fragments that showed excellent internal structure. Many characters, such as the thickness of the leaf lamina, thick cuticle, abundant fibrous tissue, sclereids and sunken stomata suggest that the plant was xerophytic. The fact that remains are often found charred supports the idea that the plant grew in dry areas where it was very vulnerable to the spread of fire.

Fossil ferns and the evolution of modern taxa are usually overshadowed in palaeobotany by studies on flowering plants. However, it is clear from the literature that a constant stream of articles has been written on fossil ferns during the last hundred years. Our understanding of both the taxonomy of fossil ferns and fern evolution has changed dramatically since Kidston's time but it is unlikely that our ideas will change so much during the next hundred years. It is anticipated that new discoveries will instead increase our general knowledge of fern fossils, which will bring with it a greater knowledge of their geological histories, ecologies, and evolution. Above all we should have a much better understanding of the origin and diversification of the early polypodiaceous ferns.

References

ALVIN, K.L. 1971. The Spore-bearing organs of the Cretaceous fern *Weichselia* Stiehler. *Bot. J. Linn. Soc.* **61**: 87-92.
---- 1974. Leaf anatomy of *Weichselia* based on fusainized material. *Palaeontology* **17**: 587-598.
ANDREWS, H.N. 1948. Fossil tree ferns of Idaho. *Archaeology* **1**: 190-195.
ARNOLD, C.A. & DAUGHERTY, L.H. 1963. The fern genus *Acrostichum* in the Eocene Clarno Formation of Oregon. *Contr. Mus. Paleont. Univ. Mich.* **18**: 205-227.
ASH, S.R. 1969. Ferns from the Chinle Formation (Upper Triassic) in the Fort Wingate area, New Mexico. *U.S. Geol. Surv. Prof. Paper* **613-D**: 1-52.
---- LITWIN R.J. & TRAVERSE A. 1982. The Upper Triassic fern *Phlebopteris smithii* (Daugherty) Arnold. *Palynology* **6**: 203- 219.
---- & TIDWELL, W.D. 1986. *Arnoldia kuesii*, a new juvenile fernlike plant from the lower Permian of New Mexico. *Bot. Gaz.* **147**: 236-242.
BECK, C.B. 1962. Reconstruction of *Archaeopteris* and further consideration of its phylogenetic position. *Amer. J. Bot.* **49**: 373-382.
BOWER, F.O. 1923-28. *The Ferns (Filicales).* vols **1-3**. Cambridge: Cambridge University Press.

BUCKLEY, D.P. 1986. *Evolution of homosporous genetic systems*. Ph.D. thesis, Ohio University, U.S.A.
COLLINSON, M.E. 1978. Dispersed fern sporangia from the British Tertiary. *Ann. Bot.* (new ser.) **42**: 233-250.
---- 1980. A new multiple-floated *Azolla* from the Eocene of Britain with a brief review of the genus. *Palaeontology* **23**: 213-229.
DIMICHELE, W.A., PHILLIPS, T.L. & PEPPERS, R.A. 1985. The influence of climate and depositional environment on the distribution and evolution of Pennsylvanian coal-swamp plants. *In* Tiffney, B.H. (ed.) *Geological factors and the evolution of plants*: 223-256. New Haven: Yale University Press.
GALTIER, J. & SCOTT, A.C. 1985. Diversification of early ferns. *Proc. R. Soc. Edinb.* **86B**: 289-301.
GAO, ZHIFENG. 1988. *Lower Permian plants from Taiyuan Shanxi Province, China*. Ph.D. Thesis, University College, Cardiff.
GREUTER, W. *et al.* 1988. *International code of botanical nomenclature*. Konigstein: Koeltz Scientific Books.
HARRIS, T.M. 1961. *The Yorkshire Jurassic flora*. **1**. *Thallophyta-Pteridophyta*. London: British Museum (Natural History).
---- 1973. What use are fossil ferns? *Bot. J. Linn. Soc.* **67** Suppl.1: 41-44.
HILL, C.R. & CAMUS, J.M. 1986. Evolutionary cladistics of marattialean ferns. *Bull. Br. Mus. nat. Hist. (Bot.)* **14**: 219-300.
---- WAGNER, R.H. & EL-KHYAL, A.A. 1985. *Qasimia* gen. nov., an early *Marattia*-like fern from the Permian of Saudi Arabia. *Scripta Geol.* **79**: 1-50.
HOLTTUM, R.E. 1982. The continuing need for more monographic studies of ferns. *Fern Gaz.* **12**: 185-190.
JAIN, R.K. 1971. Pre-Tertiary records of Salviniaceae. *Amer. J. Bot.* **58**: 487-496.
---- & HALL, J.W. 1969. A contribution to the Early Tertiary fossil record of the Salviniaceae. *Amer. J. Bot.* **56**: 527-539.
JENNINGS, J.R. & EGGERT, D.A. 1977. Preliminary report on permineralized *Senftenbergia* from the Chester series of Illinois. *Rev. Palaeobot. Palynol.* **24**: 221-225.
JOY, K.W., WILLIS, A.J. & LACEY, W.S. 1956. A rapid cellulose peel technique in palaeobotany. *Ann. Bot.* **20**: 635-637.
KIDSTON, R. 1891. On the fructification and internal structure of Carboniferous ferns in their relation to those of existing genera, with special reference to British Palaeozoic species. *Trans. geol. Soc. Glasg.* **9**: 1-56.
LACEY, W.S 1963. Palaeobotanical techniques. *Viewpts Biol.* **2**: 202-243.
LOVIS, J.D. 1977. Evolutionary patterns and processes in ferns. *In* Preston, R.D. & Woolhouse, H.W. (eds). *Advances in Botanical Research* **4**: 229-440.
MAMAY, S.H. 1950. Some American Carboniferous fern fructifications. *Ann. Missouri Bot. Gdn.* **37**: 409-459.
MICKLE, J.E. 1984. Aspects of growth and development in the Pennsylvanian age marattialean fern *Psaronius*. *Bot. Gaz.* **145**: 407-419.
MILLER, C.N. 1971. Evolution of the fern family Osmundaceae based on anatomical studies. *Contr. Mus. Paleont. Univ. Mich.* **23**: 105-169.
MORGAN, J. 1959. The morphology and anatomy of American species of the genus *Psaronius*. *Ill. Biol. Monogr.* **27**: 1-108.
NATHORST, A.G. 1906. Uber *Dictyophyllum* und *Camptopteris spiralis*. *K. svenska Vetensk Akad. Handl.* **41** pt. 5: 1-24.
OLIVER, F.W. & SCOTT, D.H. 1904. On the structure of the Palaeozoic seed *Lagenostoma lomaxi*, with a statement on the evidence upon which it is referred to *Lyginodendron*. *Phil. Trans. R. Soc. B* **197**: 193-247.
RADFORTH, N.W. 1938. An analysis and comparison of the structural features of *Dactylotheca plumosa* Artis sp. and *Senftenbergia ophiodermatica* Goeppert sp. *Trans. R. Soc. Edinb.* **59**: 385-396.
RIBBINS, M.M. & COLLINSON, M.E. 1978. Further notes on pyritised fern rachides from the London Clay. *Tert. Res.* **2**: 47-50.
ROTHWELL, G.W. 1987. Complex Paleozoic Filicales in the evolutionary radiation of ferns. *Amer. J. Bot.* **74**: 458-461.
---- & STOCKEY, R.A. 1989. Fossil Ophioglossales in the Paleocene of Western North America. *Amer. J. Bot.* **76**: 637-644.
SCOTT, A.C. 1979. The ecology of Coal Measure floras from northern Britain. *Proc. Geol. Ass.* **90**: 97-116.
SCHWEITZER, H.-J. 1978. Die Räto-Jurassischen Floren des Iran und Afghanistans. 5. *Todites princeps*, *Thaumatopteris brauniana* und *Phlebopteris polypodioides*. *Palaeontographica* B **168**: 17-60.

SEWARD, A.C. 1894. *Catalogue of the Mesozoic plants in the Department of Geology, British Museum (Natural History). The Wealden Flora, I. Thallophyta-Pteridophyta*. London: British Museum (Natural History).

---- 1900. *Catalogue of the Mesozoic plants in the Department of Geology, British Museum (Natural History). The Jurassic Flora, I. The Yorkshire coast*. London: British Museum (Natural History).

STEWART, W.S. 1983. *Paleobotany and the evolution of plants*. Cambridge: Cambridge University Press.

STIDD, B.M. 1971. Morphology and anatomy of the frond of *Psaronius*. *Palaeontographica* B **134**: 87-123.

STUBBLEFIELD, S.P. 1984. Taxonomic delimitation among Pennsylvanian marattialean fructifications. *J. Paleontol.* **58**: 793-803.

TAYLOR, T.N. 1981. *Paleobotany. An introduction to Fossil Plant Biology*. New York: McGraw-Hill.

THOMAS, B.A. 1986. The biochemical analysis of fossil plants and its use in taxonomy and systematics. *In* Spicer, R.A. & Thomas, B.A. (eds) *Systematic and taxonomic approaches in palaeobotany: Systematics Association Sp. Vol.* **31**: 39-51.

---- & SPICER, R.A. 1987. *The evolution and palaeobiology of land plants*. London: Croom Helm.

THOMAS, H.H. 1911. On the spores of some Jurassic ferns. *Proc. Camb. phil. Soc. biol. Sci.* **16**: 384-388.

TRIVETT, M.L. & ROTHWELL, G.W. 1988. Modelling the growth architecture of fossil plants: paleozoic filicalean fern. *Evolutionary Trends in Plants* **2**: 25-29.

WALTON, J. 1923. On a new method of investigating fossil plant impressions or incrustations. *Ann. Bot.* **37**: 379-391.

---- 1928. A method of preparing sections of fossil plants contained in coalballs or in other types of petrifaction. *Nature* **122**: 571-572.

Barry Thomas's early interest in plants developed into research into fossil pteridophytes, especially the Carboniferous lycopods. He has published widely on fossil and living pteridophytes, with a special interest in their evolution and palaeobiology. He held several distinguished academic posts before becoming Keeper of Botany at the National Museum of Wales in 1985. He has been involved with plant conservation at the local, national and international level for many years.

SELAGINELLA GRANDIS.

Changes in the List of British Pteridophytes

Josephine M. Camus
Department of Botany, The Natural History Museum, Cromwell Road,
London SW7 5BD

There are about 12,000 species of pteridophytes in the world but only a very few of these, 72 (at the time of writing), are native to the British Isles. A quick glance at the books written over the last two centuries about British ferns and allied plants reveals a continued history of name changes that seems quite out of proportion to the number of species. Worse, the numbers of ferns, lycopsids and horsetails have increased over the years as well: in the first book devoted solely to 'British proper ferns', James Bolton (1785-90) lists 43 species of ferns, horsetails and quillworts in 11 genera. Lowe, in 1891, described 45 species of ferns in 17 genera. One hundred years later, Jermy and Camus (1991) give the number of British species as 72 (with a further nine subspecies, morphotypes or varieties and 36 hybrids) in 30 genera.

This might imply that there has been a steady invasion of the British Isles by an assortment of pteridophytes. But this is not the case. These changes reflect the history of advances in concepts of definition and delimitation of genera and species and the relationships of these taxa to one another. The term *species* is, in simple terms, applied to a group of individuals that breed true, albeit showing morphological variation. Related species are grouped together into *genera*, and related genera are further grouped into *families*. Families are collected in *orders*. This hierarchic classification is intended to reflect, as best as possible, the evolutionary history of organisms. The concepts of genera and families were usually very broad in the past and were based on features that the members shared in common (giving paraphyletic groups), but nowadays most people try to delimit groups that are defined by features unique to the members (monophyletic groups). Although the ferns, clubmosses, quillworts and horsetails are traditionally grouped together as 'pteridophytes', these plants are not now considered to all have a common ancestry. Pteridophyta is an artificial and unnatural, paraphyletic unit.

The British pteridophytes are now arranged:

ORDERS	FAMILIES	GENERA
Lycopodiales	Lycopodiaceae	Lycopodium
		Lycopodiella
		Huperzia
		Diphasiastrum
	Selaginellaceae	Selaginella
	Isoetaceae	Isoetes
Equisetales	Equisetaceae	Equisetum
Ophioglossales	Ophioglossaceae	Botrychium
		Ophioglossum
Filicales	Osmundaceae	Osmunda
	Adiantaceae	Cryptogramma
		Anogramma
		Adiantum
	Marsileaceae	Pilularia
	Hymenophyllaceae	Hymenophyllum
		Trichomanes
	Polypodiaceae	Polypodium
	Dennstaedtiaceae	Pteridium
	Thelypteridaceae	Thelypteris
		Phegopteris
		Oreopteris

Aspleniaceae	Asplenium
Woodsiaceae	Athyrium
	Gymnocarpium
	Cystopteris
	Woodsia
Dryopteridaceae	Polystichum
	Dryopteris
Blechnaceae	Blechnum
Azollaceae	Azolla

Early taxonomists laid great weight on the shape of soral patches. Hence any ferns with round sori were called *Polypodium*; those with long sori were grouped in *Asplenium*, but if the sori covered the pinnae, the species were called *Acrostichum* or *Osmunda*; *Adiantum* and *Pteris* covered taxa with marginal sori; the filmy ferns were all called *Trichomanes*, while the adder's tongues were distinct enough to be called *Ophioglossum* by Linnaeus in 1753 when he also recognised *Pilularia*, *Isoetes* and *Equisetum* as unique groups, though he called all clubmosses *Lycopodium* and grouped them with the true mosses.

Bolton (1785-90) had followed Linnaeus as best he could, but made some mistakes in the binomial names that he applied to British species. His list of species was shorter than the one we have today largely because the concept of a species then covered a much broader range of morphology than is typical of current taxonomy. Even at that early date, Bolton was well aware that the embryonic classification of ferns was unsatisfactory as he wrote *'It must offend the taste of the judicious reader to find the characteristics of the Genera fixed on foundations so unsteady, when he finds plants very dissimilar in their appearance united; as also when he sees a separation take place between* Osmunda lunaria *and* Ophioglossum vulgatum, *between* Acrostichum septentrionale *and* Asplenium ruta-muraria.' One hundred and forty years later, Holttum (this volume) returned briefly to England after four years studying ferns in South East Asia and thought that British ferns *'were a strange-looking lot of odds and ends which appeared to need more study to connect them with the increasing studies . . . of the world's ferns as a whole.'* A prophetic thought, despite the enormous strides already made in resolving the classification of these 'odds and ends'.

Early botanists in continental Europe had the advantage of studying a wider range of species, not only from Europe but also from all over the world, and soon our British ferns bore more generic names as the first very broad groups of Linnaeus were reassessed. *Botrychium*, *Blechnum* and *Allosorus* (or *Cryptogramma*) were segregated from *Osmunda*, and the especially unwieldy *Polypodium* yielded *Gymnogramma* (now *Anogramma*), *Aspidium* or *Lastrea*, *Athyrium*, *Cystopteris* and *Polystichum*. Newman published five editions of his *A History of British Ferns* from 1840 to 1875. Throughout these he struggled with the concepts of genera and species, the problems of nomenclature (which sometimes led him to take up or invent new names and at other times to revert to better-known ones) and the variation of the plants. He discussed 18 genera with 33 species in 1840; by 1875 the number of species recorded had risen to 50 but he then dismissed the inclusion of generic names as they have *'no definite signification'* and are not used by *'a botanist really conversant with ferns'* as *'it is the invariable plan to use the specific name by itself'*. Newman includes a number of astute observations by himself and botanical correspondents and discusses, for example, the forms of the male fern that are now called *Dryopteris oreades* and *D. affinis*, and two forms of quillwort that may correspond to *Isoetes lacustris* and *I. echinospora*.

Moore published his *Popular History of British Ferns* in 1862 and summarised how complicated nomenclature had already become in his helpful lists of names by which his species were also known. Pichi Sermolli (1973) gives an excellent historical review of the different contributions made by pteridologists towards refining the classification of ferns at the family and order levels and of how the concept of a genus was eventually based on the morphology of the vegetative plant as well as that of the sorus, with many notable British names among them: J. E. Smith, R. Brown, W. J. Hooker, T. Moore, J. Smith, F. O. Bower, D. H. Campbell, R. E. Holttum, I. Manton.

For the British fern enthusiast these different lines of thought following each other in quick succession were further complicated by the staggering array of varieties and sports produced by the native ferns. Lowe (1891) described over 1,850 varieties of 45 species in 17 genera. Despite this range of morphological variants, Lowe maintained his and Wollaston's (1875) earlier concepts that the common male fern, *Dryopteris filix-mas*, was really the three species that we now know as *D. filix-mas*, *D. oreades* and *D. affinis*. In 1939, Manton showed that they had different chromosome numbers. Fraser-Jenkins (in press) has further elaborated the complexities of *D. affinis*.

Dryopteris is a good example of the evolution of our changing concept of a species. Bolton, towards the end of the 18th century, had only three names for the now ten British species of this genus. Whilst one species, *Dryopteris remota*, is an apomict of hybrid origin and has only been found twice in the British Isles, the other nine species were without doubt growing here in Bolton's time. Linnaeus' *Polypodium cristatum* was an umbrella for five species; different, but with a complex ancestry of shared genomes unravelled in the 20th century by the study of their chromosomes and chemicals. Gibby (this volume) reviews the development of these fields which have been especially significant in studies of *Polypodium* and *Asplenium* as well as *Dryopteris*.

The six lists given here, from just a few of the many books on British ferns, show the problems of generic limits. *Acrostichum thelypteris*, *Polypodium phegopteris* and *Aspidium oreopteris* have been shuffled through a number of genera before being settled in the family Thelypteridaceae as *Thelypteris palustris*, *Phegopteris connectilis* and *Oreopteris limbosperma* respectively. The 26 synonyms of the latter given in Derrick *et al.* (1987) are provocative of such comments as Druery's (1910) in the introduction to *British Ferns and their Varieties*: *'The generic and specific names given are also those generally recognised by British Fern-growers, and we have purposely steered clear of the terrible quagmire involved in the infinite number of synonyms, or different names for the same thing, resulting from varied and frequently mistaken views on the part of botanists who make classification and nomenclature their study, many of whom, too, are constantly inventing new names for old friends, and thus turning confusion into chaos.'* It was all too easy for a botanist studying his local ferns, which might well display the opposite extreme of morphological variation to those ferns of the same species studied by another botanist in another country, perhaps on the opposite side of the world, to think that he had a new species. However, another factor is the International Code of Botanical Nomenclature, drawn up and continually revised by international panels of taxonomists, which stipulates that except in exceptional cases (and alas, *Polypodium australe* was not amongst these and has had to be called *P. cambricum*), the earliest valid name must be taken up. The more thorough a modern taxonomist is in his or her researches, especially into temperate plants, the more names are likely to be turned up. Hence a novice on a Society's field meeting may well be flummoxed by hearing members of different ages using different specific epithets such as *rigida*, *villarii* or *submontana* for the same plant.

Ferns with apogamous reproductive cycles (apomicts) or with different ploidy levels (multiples of the basic chromosome numbers) presented something of a problem as regards rank, but Lovis (1981) devised the policy of treating autopolyploids (plants where one set of chromosomes has doubled) as subspecies and allopolyploids (plants where two sets of chromosomes are involved) as species.

Individual opinion also has its role. The hart's tongue fern (*Asplenium scolopendrium*) has been variously considered as a species of *Asplenium* or worthy of separate generic status as *Phyllitis*. The discovery of its naturally occurring hybrids with *Asplenium*

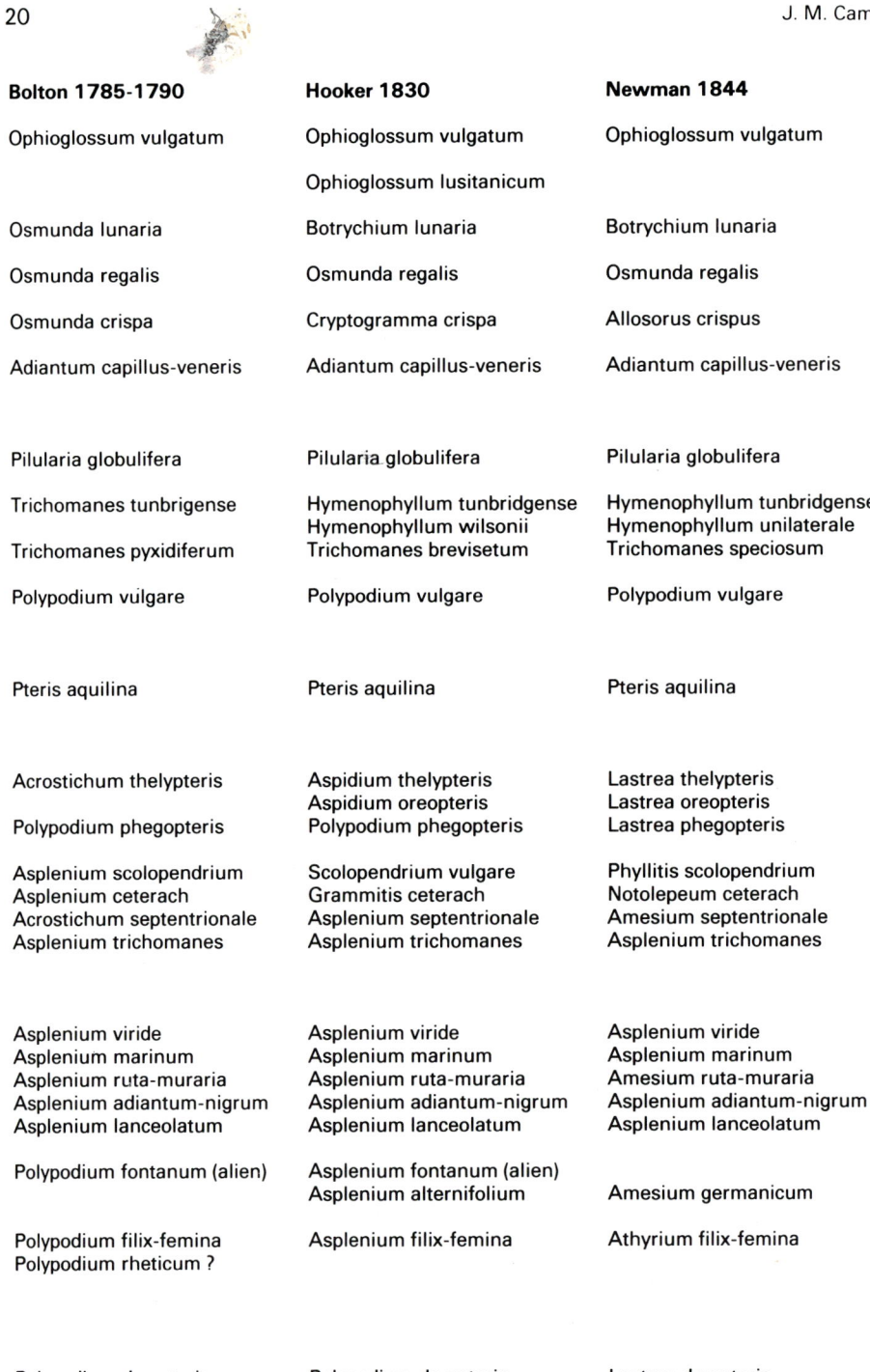

Bolton 1785-1790	Hooker 1830	Newman 1844
Ophioglossum vulgatum	Ophioglossum vulgatum	Ophioglossum vulgatum
	Ophioglossum lusitanicum	
Osmunda lunaria	Botrychium lunaria	Botrychium lunaria
Osmunda regalis	Osmunda regalis	Osmunda regalis
Osmunda crispa	Cryptogramma crispa	Allosorus crispus
Adiantum capillus-veneris	Adiantum capillus-veneris	Adiantum capillus-veneris
Pilularia globulifera	Pilularia globulifera	Pilularia globulifera
Trichomanes tunbrigense	Hymenophyllum tunbridgense	Hymenophyllum tunbridgense
	Hymenophyllum wilsonii	Hymenophyllum unilaterale
Trichomanes pyxidiferum	Trichomanes brevisetum	Trichomanes speciosum
Polypodium vulgare	Polypodium vulgare	Polypodium vulgare
Pteris aquilina	Pteris aquilina	Pteris aquilina
Acrostichum thelypteris	Aspidium thelypteris	Lastrea thelypteris
	Aspidium oreopteris	Lastrea oreopteris
Polypodium phegopteris	Polypodium phegopteris	Lastrea phegopteris
Asplenium scolopendrium	Scolopendrium vulgare	Phyllitis scolopendrium
Asplenium ceterach	Grammitis ceterach	Notolepeum ceterach
Acrostichum septentrionale	Asplenium septentrionale	Amesium septentrionale
Asplenium trichomanes	Asplenium trichomanes	Asplenium trichomanes
Asplenium viride	Asplenium viride	Asplenium viride
Asplenium marinum	Asplenium marinum	Asplenium marinum
Asplenium ruta-muraria	Asplenium ruta-muraria	Amesium ruta-muraria
Asplenium adiantum-nigrum	Asplenium adiantum-nigrum	Asplenium adiantum-nigrum
Asplenium lanceolatum	Asplenium lanceolatum	Asplenium lanceolatum
Polypodium fontanum (alien)	Asplenium fontanum (alien)	
	Asplenium alternifolium	Amesium germanicum
Polypodium filix-femina	Asplenium filix-femina	Athyrium filix-femina
Polypodium rheticum ?		
Polypodium dryopteris	Polypodium dryopteris	Lastrea dryopteris
	Polypodium calcareum	Lastrea robertiana
Polypodium fragile	Cistopteris fragilis	Cystopteris fragilis
	Cistopteris dentata	
	Cistopteris alpina	
		Cystopteris montana

Lowe 1891	Hyde & Wade 1940	Jermy & Camus 1991
Ophioglossum vulgatum	Ophioglossum vulgatum	Ophioglossum vulgatum
		Ophioglossum azoricum
Ophioglossum lusitanicum	Ophioglossum lusitanicum	Ophioglossum lusitanicum
Botrychium lunaria	Botrychium lunaria	Botrychium lunaria
Osmunda regalis	Osmunda regalis	Osmunda regalis
Cryptogramma crispa	Cryptogramme crispa	Cryptogramma crispa
Adiantum capillus-veneris	Adiantum capillus-veneris	Adiantum capillus-veneris
Gymnogramma leptophylla	Anogramma leptophylla	Anogramma leptophylla
	Pilularia globulifera	Pilularia globulifera
Hymenophyllum tunbridgense	Hymenophyllum tunbrigense	Hymenophyllum tunbrigense
Hymenophyllum unilaterale	Hymenophyllum peltatum	Hymenophyllum wilsonii
Trichomanes radicans	Trichomanes speciosum	Trichomanes speciosum
Polypodium vulgare	Polypodium vulgare	Polypodium vulgare
		Polypodium interjectum
		Polypodium cambricum
Pteris aquilina	Pteridium aquilinum	Pteridium aquilinum
		morphotype aquilinum
		morphotype latiusculum
Nephrodium thelypteris	Thelypteris palustris	Thelypteris palustris
Nephrodium montanum	Thelypteris oreopteris	Oreopteris limbosperma
Polypodium phegopteris	Thelypteris phegopteris	Phegopteris connectilis
Scolopendrium vulgare	Phyllitis scolopendrium	Asplenium scolopendrium
Asplenium ceterach	Ceterach officinarum	Asplenium ceterach
Asplenium septentrionale	Asplenium septentrionale	Asplenium septentrionale
Asplenium trichomanes	Asplenium trichomanes	Asplenium trichomanes
		ssp. trichomanes
		ssp. quadrivalens
		ssp. pachyrachis
Asplenium viride	Asplenium viride	Asplenium trichomanes-ramosum
Asplenium marinum	Asplenium marinum	Asplenium marinum
Asplenium ruta-muraria	Asplenium ruta-muraria	Asplenium ruta-muraria
Asplenium adiantum-nigrum	Asplenium adiantum-nigrum	Asplenium adiantum-nigrum
Asplenium billotii	Asplenium obovatum	Asplenium obovatum
		ssp. lanceolatum
Asplenium fontanum (alien)		
Asplenium germanicum	Asplenium breynii	Asplenium x alternifolium
Asplenium filix-femina	Athyrium filix-femina	Athyrium filix-femina
		Athyrium distentifolium
Polypodium alpestre	Athyrium alpestre	var. distentifolium
	Athyrium flexile	var. flexile
Polypodium dryopteris	Gymnocarpium dryopteris	Gymnocarpium dryopteris
Polypodium robertianum	Gymnocarpium robertianum	Gymnocarpium robertianum
Cystopteris fragilis	Cystopteris fragilis	Cystopteris fragilis
Cystopteris alpina	Cystopteris alpina	
	Cystopteris dickieana	Cystopteris dickieana
Cystopteris montana	Cystopteris montana	Cystopteris montana

Bolton *cont.*	Hooker *cont.*	Newman *cont.*
Acrostichum ilvense	Woodsia ilvensis	Woodsia ilvensis
Acrostichum alpinum	Woodsia alpina	Woodsia alpina
Polypodium aculeatum	Aspidium aculeatum	Polystichum aculeatum
Polypodium lobatum	Aspidium lobatum	Polystichum angulare
	Aspidium angulare	
Polypodium lonchitis	Aspidium lonchitis	Polystichum lonchitis
Polypodium cristatum	Aspidium dumetorum	Lastrea multiflora
		Lastrea recurva
	Aspidium spinulosum	Lastrea spinosa
		Lastrea rigida
Polypodium thelypteris	Aspidium cristata	Lastrea callipteris
Polypodium filix-mas	Aspidium filix-mas	Lastrea filix-mas
Osmunda spicant	Blechnum boreale	Lomaria spicant
	Lycopodium selago	Lycopodium selago
	Lycopodium annotinum	Lycopodium annotinum
	Lycopodium clavatum	Lycopodium clavatum
	Lycopodium inundatum	Lycopodium inundatum
	Lycopodium alpinum	Lycopodium alpinum
	Lycopodium selaginoides	Lycopodium selaginoides
Isoetes lacustris	Isoetes lacustris	Isoetes lacustris
Equisetum arvense	Equisetum arvense	Equisetum arvense
Equisetum hyemale	Equisetum hyemale	Equisetum hymale
Equisetum limosum	Equisetum limosum	Equisetum fluviatile
		Equisetum mackaii
Equisetum fluviatile	Equisetum fluviatile	Equisetum telmateia
Equisetum palustre	Equisetum palustre	Equisetum palustre
Equisetum sylvaticum	Equisetum sylvaticum	Equisetum sylvaticum
	Equisetum drummondii	Equisetum umbrosum
	Equisetum variegatum	Equisetum variegatum

Lowe *cont.*	Hyde & Wade *cont.*	Jermy & Camus *cont.*
Woodsia ilvensis	Woodsia ilvensis	Woodsia ilvensis
Woodsia hyperborea	Woodsia alpina	Woodsia alpina
Aspidium aculeatum	Polystichum aculeatum	Polystichum aculeatum
Aspidium angulare	Polystichum setiferum	Polystichum setiferum
Aspidium lonchitis	Polystichum lonchitis	Polystichum lonchitis
Nephrodium spinulosum	Dryopteris dilatata	Dryopteris dilatata
		Dryopteris expansa
	Dryopteris aemula	Dryopteris aemula
	Dryopteris spinulosa	Dryopteris carthusiana
Nephrodium rigida	Dryopteris villarsii	Dryopteris submontana
		Dryopteris remota
Nephrodium cristata	Dryopteris cristata	Dryopteris cristata
Nephrodium filix-mas	Dryopteris filix-mas	Dryopteris filix-mas
Nephrodium propinquum	var. abbreviata	Dryopteris oreades
Nephrodium paleaceum	var. paleacea	Dryopteris affinis
		morphotype affinis
		morphotype palaeceo-lobata
		morphotype borreri
		morphotype robusta
		morphotype cambrensis
Lomaria spicant	Blechnum spicant	Blechnum spicant
	Azolla filiculoides	Azolla filiculoides
		Huperzia selago
		Lycopodium annotinum
		Lycopodium clavatum
		Lycopodiella inundata
		Diphasiastrum alpinum
		Diphasiastrum complanatum
		morphotype decipiens
		Selaginella selaginoides
		Isoetes lacustris
		Isoetes echinospora
		Isoetes histrix
		Equisetum arvense
		Equisetum hyemale
		Equisetum fluviatile
		Equisetum x trachyodon
		Equisetum telmateia
		Equisetum palustre
		Equisetum sylvaticum
		Equisetum pratense
		Equisetum ramossisimum
		Equisetum variegatum

trichomanes ssp. *quadrivalens* and *A. billotii* adds weight to the argument of those who regard it as part of *Asplenium*, but some authors such as Page (1982, 1988) and Lovis (in Stace, 1975) have treated it as a separate genus. The treatment reflects the individual's assessment of how fundamental are differences, such as those seen in the indusia in this example.

There will be more changes to the list of ferns, lycopsids and horsetails of the British Isles. Derrick *et al.* (1987) already merits a revised version. More discoveries await us in new hybrids, subspecies, and naturalised or adventive species. More information will come from studies of chromosomes and chemicals. Meanwhile, Jermy and Paul are preparing a checklist of the pteridophytes of the British Isles that will include all synonyms and hybrids with their parentage and is scheduled for publication in 1991.

References

BOLTON, J. 1785-1790. *Filices Britannicae*. Vol. 1 Leeds: Binns; Vol.2 Huddersfield: Brook.
DERRICK, L.N., JERMY, A.C. & PAUL, A.M. 1987. Checklist of European pteridophytes. *Sommerfeldtia* **6**: i-xx, 1-94.
DRUERY, C.T. 1910. *British ferns and their varieties*. London: Routledge.
GIBBY, M. 1991. The development of laboratory based studies in fern variation. *In* Camus, J.M. (ed.), *The history of British pteridology 1891-1991*: 59-63. London: British Pteridological Society.
HOLTTUM, R.E. 1991. My life with ferns. *In* Camus, J.M. (ed.), *The history of British pteridology 1891-1991*: 41-43. London: British Pteridological Society.
HOOKER, W.J. 1830. *The British flora*. London: Longman *et al.*
HYDE, H.A. & WADE, A.E. 1940 (1st ed.) *Welsh ferns*. Cardiff: National Museum of Wales.
JERMY, A.C. & CAMUS, J.M. 1991. *The illustrated field guide to ferns and allied plants of the British Isles*. London: Natural History Museum Publications.
LINNAEUS, C. 1753. *Species plantarum*. Stockholm: Salvi.
LOVIS, J.D. 1975. *Asplenium*. *In* Stace, C. (ed.), *Hybridisation in the British Isles*: 106-112. London: Academic Press.
---- 1981. Hybrids in European Aspleniaceae (Pteridophyta). *Botanica Helvetica* **91**: 89-139.
LOWE, E.J. 1891. *British ferns*. London: Swan Sonnenschein & Co.
MANTON, I. 1939. Cytology of the common male fern in Britain. *Nature* **144**: 291.
MOORE, T. 1862. *A popular history of British ferns*. London: Routledge, Warne.
NEWMAN, E. 1840, 1844, 1854, 1874. *A history of British ferns*. London: van Voorst.
PAGE, C.N. 1982. *The ferns of Britain and Ireland*. Cambridge: Cambridge University Press.
---- 1988. *Ferns*. London: Collins New Naturalist.
PICHI SERMOLLI, R.E.G. 1973. Historical review of the higher classification of the Filicopsida. *In* Jermy, A.C., Crabbe, J.A. & Thomas, B.A. (eds), *The phylogeny and classification of the ferns*. *Bot. J. Linn. Soc.* **67** suppl. 1: 11-40.
WOLLASTON, G.B. 1875. Three species of *Lastrea filix-mas*. *Occ. paper Brit. Pterid. Soc.* **1**: 9-13.

Josephine Camus is following in the footsteps of her predecessors at The Natural History Museum, London, with her main research interest in tropical pteridophytes. She enjoys gardening but prefers to be with ferns in the wild rather than to cultivate fancy ones!

Pteridophytes as Indicators of Landscape Changes in the British Isles in the Last Hundred Years

C. N. Page and H. S. McHaffie
Royal Botanic Garden, Edinburgh, Scotland EH3 5LR

Introduction

The pteridophyte flora of the British Isles, like that of its flowering plants (e.g. Walters, 1984), is small in comparison with that of Europe as a whole. But as with the bryophytes of these islands (Ratcliffe, 1968), it contains many unusually Atlantic elements within a European context, setting a theatrical stage into which more local species as well as an unusually large number of hybrids (Jermy et al., 1978; Page, 1982b) have become important players within a European context.

The ways in which the major features of our overall landscape and vegetation have been influenced by man in Britain and Ireland, both before and during the past century, have been extensively studied (e.g. Hoskins, 1970; Godwin, 1975; Mitchell, 1976; Webb, 1983; Dimbleby, 1978, 1984; Rackham, 1980, 1986; Ratcliffe 1984). These studies provide a particularly valuable background against which to set an interpretation of the ways in which the pteridophytes have responded, in our insular environment, to the general habitat changes resulting (Page, 1988).

Within the last century, the changes which have affected the landscapes of the British Isles, and thereby influenced the habitats available to pteridophytes, have been many and various. Those discussed here are some of the more major of these, whose influences have been felt widely in these islands, and which have served either to destroy or modify previous pteridophyte habitats, or to gain a suite of new ones.

Even in a flora so small as that of these islands, however, there is, among the pteridophytes, a considerable but scarcely-appreciated potential indicator value to many specific facets of environmental change, the study of which provides a valuable scientific basis for integration into broader aspects of conservational issues and towards the formulation of future management plans for many of the environments and locations around us.

Changes in forest extent and type

A steady reduction in the amount of natural forest cover in the British Isles has proceeded ever since man first set foot in these islands, but perhaps never more rapidly than in the period under consideration. This increasing rate of loss of remaining natural and semi-natural forest areas through the last century has continued to extensively reduce the original habitats for many woodland ferns, not only by direct removal of tree cover, but also by drainage of the majority of areas of originally naturally wet woodland. Today, some of the best remaining semi-natural woodland habitats for pteridophytes persist especially along steep stream and river ravines, while flatter areas of wet woodland have severely diminished, of which the modern scarcity of narrow buckler fern (*Dryopteris carthusiana*) provides a particularly vivid indicator.

In contrast to the original diversity of native forest vegetation, within the modern uplands and on poorer soils at low altitude, extensive forestry plantations, typically of conifers such as Sitka spruce, have provided a new standardised forest of the 20th century. In these, trees are treated as crops, and are typically clear-felled on a rotational basis of about 40 years, from whence the cycle begins again. Few native plants either tolerate this cycle or successfully compete with the high and even-aged density with which the spruce is artificially set. The resulting occurrence of pteridophytes in these communities is seldom little more than the limited success of a few of the commoner woodland ferns along marginal ditches and rides, or the persistence of streamside communities along ravines, where a break in the dense and monotonous tree canopy allows necessary extra light.

In most cases, such forests have been responsible for loss rather than gain of pteridophyte biodiversity, especially for those species of natural wet heathlands. A particular pteridophyte indicator of such losses may well be marsh clubmoss (*Lycopodiella inundata*), which has almost certainly suffered from afforestation extending over its sites. In 1844 William Gawlee wrote on a herbarium specimen of this species (Glasgow University Botany Department) from the Carse of Arderries: *'Two miles east of Fort George is a moor now planted with wood, which will in a few years, I fear, make a radical reform in the Vegetable Kingdom of this locality'*. Webster (1978) recorded the similar demise of *Lycopodiella* from the Culbin sands. Earlier writers (Patton & Stewart, 1914, 1924) had recorded its presence in the dune slack areas but Webster saw it last in 1976 where it grew in seasonally inundated hollows, which had become more shaded and drier with tree upgrowth and forestry drainage.

Where rocks emerge in forests, small enclaves of existing sun-loving ferns are especially threatened by rapid tree upgrowth from dense conifer planting. There are also at least two sites in Scotland and one in Devon, for example, in which forked spleenwort (*Asplenium septentrionale*) is critically diminishing (in two of them already extinguished) through forestry planting (of Douglas fir in both Scottish sites, and Sitka spruce in Devon). In each of these cases, greater sensitivity to the existence of such already rare and localised species in initial forestry planting could well have saved them, at the expense, at each site, of only a handful of trees.

From the pteridophyte point of view, a conservationally more sensitive forest management policy for Britain would certainly be in not only awareness when planting of existing rare native species, but also of the overall progressive future construction of alternative forest types altogether. Conservationally infinitely more valuable would be ones of intermixed broadleaf and evergreen communities with wide planting-spaces and uneven-aged structures (mimicking the natural form of the majority of northern hemisphere forests of high latitudes), managing these forests on a principle of permanency of tree-cover, high contained and edge-effect habitat diversity, and sustained yield of more mature and more unusual timbers. In comparison with existing plantation tree-farms, such alternative forests have already successfully long existed, in limited extent, in the much more enlightened woodland plantings policy of many larger estates. Pteridophyte study in these habitats by the authors shows that, even when well-established and although artificial, such forest types do more typically come to contain a relatively rich assemblage of native fern species, including those of wetter habitats. They can also include more unusual elements, providing forest refuges for horsetails (notably wood horsetail, *Equisetum sylvaticum* [from Perthshire to Hampshire], but also shade horsetail, *E. pratense* in northern sites), as well as even for clubmosses (notably stag's-horn clubmoss, *Lycopodium clavatum* and alpine clubmoss, *Diphasiastrum alpinum*, in northern and western Britain) and interrupted clubmoss, *Lycopodium annotinum*, in at least one policy plantation pinewood in Perthshire. In these woodlands, the presence of woodland ferns, woodland horsetails and clubmosses, as well as the overall diversity of woodland flowering plants, mammals, insects and birds, provides particularly good indicators of the conservational value and status and potential for restoration of such managed woodland vegetation, in comparison with present nationally widespread tree-farm forests, from which conversion to such vegetation is yet possible.

Wetland drainage, losses of ponds, and changes in reservoir and lake-surface levels

The changes which have been wrought this century to wetland areas and standing waters of all kinds seem to be enormous, and have widely affected the few pteridophyte species associated with such habitats.

Wetland drainage

Drainage of natural wetland areas has occurred in lower-lying districts of the British landscape, either for gain of agricultural land or for peat extraction or for both, and has certainly widely affected some wetland pteridophyte frequency through wholesale destruction of entire habitats. Although the beginnings of this drainage long pre-date the last hundred years, nevertheless throughout this century demand for land and availability of appropriate machinery has enabled the process to continue with rapidly increasing momentum. Especially pteridologically affected have been species-rich fenlands in regions of Britain as far apart as south-western Scotland, Yorkshire, East Anglia, and the Somerset levels, and the rates of resulting losses of especially royal fern (*Osmunda regalis*), fen buckler-fern (*Dryopteris cristata*) and marsh fern (*Thelypteris palustris*) provide particularly apt pteridological indicators.

Ponds, reservoirs and lake-level changes

At the smaller end of the scale, ponds of all sizes have disappeared from farmland all over Britain, especially in post-1945 years, the speed of such processes probably stimulated by available agricultural grant-assistance. The pond losses have been ones largely of lowland Britain, with a relatively rapid loss of many muddy marginal habitats and consequent loss of the marginal plant species (as well as animal populations) dependent upon them. In some areas, however, notably in the Tertiary deposits of south and south-east England, new areas of water have also been created by flooding of former gravel extraction workings.

At the larger end of the scale, numerous old lake and former stream valleys have disappeared under new reservoir schemes, often destroying a whole diversity of pteridophyte stream-valley habitats beneath them. The new water levels, with initially often steep and rocky shorelines, by comparison are usually rather sterile pteridophyte sites, which have often only now scarcely begun to carve their own new habitats. The changes of the natural lake levels of the central Scottish Highlands by Scottish hydro-electric schemes of this century provide particularly vivid examples of quantifiable change. In 1950 at Loch Tummel, Perthshire, the level was raised by 17 feet (5.1 m), making its extent two miles (3.2 km) longer. The level of Loch Lyon was raised by 70 feet (21 m). Lochan Breaclaich was raised by 66 feet (nearly 20 m) and a completely new loch was created in Glen Lednoch. Lochan na Lairige was enlarged with a conspicuous dam between Ben Lawers and Meal nan Tarmachan. New lochs were created like Dunalistair, Loch Errochty and Loch Faskally while Loch Giorra and Loch Daimh were dammed and combined. There are nine power stations on the River Tummel and tributaries alone and other rivers have similar complexes (Taylor, 1979). In contrast to ponds, such reservoir changes are ones that largely (though not exclusively) change habitats of upland Britain, and similar examples can be found in the Pennines, central Wales and south-west England.

All of these changes have had both negative and positive effects on pteridophytes, with some of the losses partly offset by changes in range of other aquatic pteridophytes, including the horsetails and quillworts. Around larger bodies of water especially, the horsetails (*Equisetum* spp.) and their hybrids provide particularly useful indicators of change. Temporary low levels in a reservoir can provide sites for the germination of *Equisetum* spores (Page, 1967), and the occurrence of different shoreline hybrid horsetails can provide good pteridophyte indicators of different degrees and periodicity of shoreline disturbance. Quillworts (*Isoetes* spp., notably *I. echinospora*) may have increased their overall range, perhaps carried as spores to new reservoirs on the feathers of migrating water fowl (Page, 1982a). But for many such landscape modifications, a particularly sensitive pteridophyte indicator of change seems to be the pillwort, *Pilularia globulifera*.

In the last hundred years in the lowlands, *Pilularia globulifera* has widely disappeared, surviving mainly only in districts where ponds have persisted more widely, such as in the Hampshire basin. The species also seems to have shown particularly wide-range changes in Scotland, disappearing in many smaller ponds but appearing in some of the newer water areas created, where its appearance is often associated with the steady re-accumulation of new lakehead silt habitats resulting from abrupt alterations in the original water level around new reservoirs and lochs. The water level of Loch Tummel, for example, was raised in 1950, and *Pilularia* was recorded from the new lakehead habitats evolving in 1965 (and has persisted since), when it had not previously been known in the loch. It also occurs in deeper water off-shore from the main colony. This seems to be the mode of growth which has been adopted in several of the deeper lochs, while in naturally shallower ones, it has sometimes formed more extensive carpets. In Kirkcudbrightshire, for example, when Jordieland Loch was drained for maintenance in 1960, the shore was found to be abundantly covered with *Pilularia*. Loch Ken in the same area, which was formed by damming the river, was found under similar circumstances to have *Pilularia* growing on the former river banks which had been inundated (Stewart, 1988).

Agricultural landscape changes

Throughout the period under consideration, there has been a very general tendency for a landscape change from smaller, more diversified farms, to fewer, larger, more intensive ones, widely across Britain (less so in Ireland), and this trend has been increasingly heightened in the post Second World War period through to today. Some of the most obvious of these changes are in the widespread losses of ponds (see above) and the similar losses of woodland and hedgerow habitats (the latter themselves miniature remnant woodland environments) and their marginal zones, which have accompanied these changes (e.g. Clapham, 1953; Hooper, 1970, 1974). Other less obvious ways in which pteridophytes have been influenced by steady agricultural intensification have been through ploughing of old pastureland and water-meadow sites; in subtle changes in grazing regimes which have had particularly profound effects on pteridophyte survival; and in the practice of moorland burning ('muirburn' in Scotland). Because of the large area of the land surface of Britain devoted to agriculture (78% in 1978, Poore, 1985:191), these processes have often had particularly widespread effects on the diminishing survival of an exceptional range of pteridophyte communities, while the effect of grazing regimes and moorland burning both spread well beyond that of land classified as agricultural.

Losses of old pasture

According to Poore (1985), of the 78% of Britain's land surface devoted to agriculture, 48% is grassland. The pteridological importance in these islands of the turf of old, semi-natural pasture grasslands is that, in appropriate sites, these are the most ideal (and over large areas, almost the exclusive) habitats for the success of especially adder's tongue (*Ophioglossum vulgatum*) and moonwort (*Botrychium lunaria*). Both of these ancient ferns seem to have been widely known as regular and sometimes abundant members of grassland communities in the lowlands of Medieval England (Page, 1988), and both of these plants have certainly widely disappeared from many of their former sites.

Ophioglossum vulgatum still survives, however, in undisturbed sites, and its present habitats are probably mostly those which have persisted over a very long period of human history in these islands. It is especially typical of old, level, low-lying, unploughed, moist, grassy meadows, developed over deep, often heavy and usually markedly basic soils, and of old, moist, water-meadows and pastures in the rural lowlands of central and southern England, which have evolved a good species diversity. It is a species of the slight hollows which fill with rain, and it can consequently sometimes be characteristic of the damper, greener turf of the furrows in old ridge-and-furrow pasture. By contrast, *Botrychium lunaria* is a less gregarious species than adder's tongue, seldom forming such dense or numerous colonies, and occurring more often as rather more scattered individuals throughout lightly-grazed, grassland turf. In both genera, the materials supplied by the mycorrhizal fungal associates are probably largely derived from the steady soil breakdown of dead organic material deriving from many of the other plants in the neighbouring grassland sward, perhaps especially the grasses. Such steady states of dynamic equilibrium are probably built up only gradually over long periods of time, but can be rapidly destroyed by sudden external disturbance.

Both *Ophioglossum* and *Botrychium* are consequently particularly easily killed in agricultural grassland by break-up of their colonial structure, and typically catastrophically so by sudden ploughing, herbicide treatment and re-seeding of old meadow and pasture grasslands, although these species, and especially Ophioglossum, seem able to persist through slower, more natural, grassland changes. To judge from 19th century flora narratives and herbarium specimens, very much of this loss, of which these two plants are probably among the most sensitive indicators, has occurred mainly through the last century.

Intensity of grazing pressures

Increasing intensity of agricultural grazing pressures has also heavily influenced the distributions of many pteridophytes, in a considerable range of habitats. This includes not only the more managed grazing by cows and sheep, but also that of fluctuating rabbit populations and, in appropriate (especially upland) areas, that of increasing populations of deer. None of these animals are new, of course, but over the last century, the factor which has changed is, in most cases, one of degree (e.g. Hunter, 1962).

Many pteridophytes appear to provide especially sensitive and hence valuable indicators of such short- and long-term intensity of grazing pressures, occasionally in positive but more usually in negative ways, which can be especially seen in habitats which are marginal to lowland agriculture, and thereby act as important refuges for pteridophyte diversity. For although ferns may not seem to be very palatable, their susceptibility in spring when flushing tender, succulent croziers for the whole season's fronds is especially great. Such times coincide with a season when other forage may be at a premium.

Further, the evergreen ferns and the evergreen horsetails (notably dutch rush, *Equisetum hyemale*, but also variegated horsetail, *E. variegatum*) appear to be grazed heavily in winter when other vegetation is also scarce. Our observations show that in captivity, rabbits, for example, can eagerly seek and totally consume the shoots of *E. hyemale* and *E. variegatum*) in winter, to a degree which suggests that these might be important items of their diet when available in the field. Field evidence indicates that *E. hyemale* is also consumed out of existence by cattle grazing, wherever such stock has the opportunity to eat it (Page, 1988), and presumably so also, in more upland areas, by the much more omnivorous deer. Much of the halving of the abundance of *E. hyemale* in Britain in the last hundred years, especially throughout agricultural counties (see Jermy *et al.*, 1978:14), we attribute largely to exposure to increasing pressure of grazing of all sorts. Personal observations have also shown that water fowl may browse horsetails, notably water horsetail (*E. fluviatile*) in the nesting season, and recent evidence from ornithological research has also come to the fore that independently suggests that horsetails form an important source of food for migrating geese, these birds obtaining an energy-important fatty acid (linoleic acid) essential during migration flights, as well as a valuable source of protein, from these ancient plants (Thomas & George, 1975; Thomas & Prevett, 1982; Fox, pers. comm. 1990).

In the field, effects of grazing are always insidious, and hence the absence of a plant through grazing pressure is clearly difficult to prove. Such grazing pressures over the last century have almost certainly had especially widespread long-term effects on our native woodlands (e.g. Morgan, 1936; Peterken, 1974; Perring, 1974; Rackham, 1976; 1986), as well as upland vegetation (e.g. Tansley, 1939, 1949; Pearsall, 1950; Raven

& Walters, 1956; McVean & Ratcliffe, 1962; Gimmingham, 1964; King & Nicholson, 1964; Ratcliffe, 1977; Sydes & Miller, 1988; Mardon, 1990), and especially on the pteridophyte species within each of these major vegetation units. Many specific examples of recovery in pteridophyte abundance with sudden reversal in grazing regimes do, however, exist, which certainly show direct cause-and-effect linkages – in the case of rabbit-grazing, especially following the abrupt and widespread decimation of rabbit populations in the late 1950s through the outbreak of myxomatosis (Thomas, 1963).

Especially in coastal and other unploughed lowland vegetation, all three species of native *Ophioglossum* provide examples of good pteridophyte indicators of grazing regimes. In these cases, under low grazing pressure, all may be helped to succeed by removal of competing upgrowth. A particular case in point here is the loss of sites for *Ophioglossum lusitanicum* in the Channel Islands (McClintock, 1975) through the alternative upgrowth of invading gorse. Further Rackham (1986) described the brief luxuriance of *Botrychium* in Breckland after the demise of the rabbits, but as taller grasses grew more successfully the *Botrychium* suffered from excessive shading. Further, in a study of rabbit-free vegetation in Norfolk, White (1961) reported that *Polypodium vulgare* had spread markedly on old dunes since the rabbits departed. Personal observations have shown too, in the presence of grazing on the Kirkudbrightshire coast, that this latter species flourishes especially on the mounds created by active nests of ants, which appear if disturbed, and it seems possible that the plant here gains a grazing defence against herbivore depredations by its associations with such sites. These are instances of probably much more widespread phenomena in which, in some habitats, light grazing pressure may maintain a critical ecological balance, whereas in others any grazing may be harmful, and clearly much more research is yet needed.

In more upland and inland woodland areas too, high differential susceptibility of different pteridophytes to different grazing pressures, appears to exist. There is extremely little literature on this, but we believe it to be an extremely profound and widespread phenomenon, in which the potential indicator value of pteridophytes has been little appreciated. Independent observations by both authors in Highland Scotland, but especially in Perthshire, for example, have shown that such upland species as *Dryopteris affinis* can be especially heavily and extensively grazed in some seasons, probably by both roe and red deer. Other species seem to be less attractive, but *Athyrium distentifolium* [var. *distentifolium* Ed.] appears to be another extremely grazing-susceptible species. It has previously, for example, also been described (from Caenlochan) as a fern which had been *'much disfigured through being eaten down by the deer'* (Cowan, 1911:173), and personal observations suggest that this can be so in many of its sites. As well, *Oreopteris* and *Polystichum* are particularly frequently browsed by deer, the latter especially throughout the winter months. Such grazing factors are probably extremely important ones in limiting surviving populations especially to grazing-inaccessible ledges or under boulders (especially all upland *Athyrium*) and to deep holes in block scree (especially both upland species of *Polystichum*) (Ratcliffe, 1977; Page, 1988).

In other native woodland habitats, personal observations also suggest widespread changes in pteridophyte vegetation to have taken place especially in the last century as a result of grazing pressures of larger herbivores, and that this is true of different animal pressures in many types of woodland and forest (*cf.* also Peterken, 1981). In native pinewood vegetation, for example, in many areas of Speyside, where deer grazing is especially heavy, almost all ferns (except bracken) are notably absent over large areas in which appropriate habitats are manifold, and these areas appear to coincide with ones where winter grazing of large populations of deer is especially heavy. Experimental grazing-exclosure areas show how both fern regeneration (as well as that of pine and often juniper) is held constantly in check by such grazing pressure. Once excluded, all these plants recover. In native oak woodland vegetation too, personal observations show a similar picture. In south-west Irish oak wood vegetation known to be one in which atlantic species of pteridophytes can potentially thrive in particular abundance (e.g. Kelly, 1981), personal observations show that today, apart from bracken, there is a surprising dearth of common woodland ferns through substantial areas of these woods. Within limited grazing-exclosure sites established for experimental purposes, however, there is by contrast vigorous regeneration of those ferns which are largely absent in the grazed areas. These include especially hard fern (*Blechnum spicant*), hay-scented buckler-fern (*Dryopteris aemula*) and woodland lady-fern (*Athyrium filix-femina*). Where grazing is excluded, furthermore, there develops also a good natural suppression of invading bracken by seedling tree upgrowth. Almost certainly, a similar general suppression of fern growth is today apparent, but scarcely appreciated, in many of our woodlands, including especially virtually all of our Royal Forests, with consequent spread of bracken and steady depletion of their pteridophyte biodiversity.

Indeed, bracken, *Pteridium aquilinum*, has spread extensively in all types of marginal land extensively in the last century, almost certainly largely as a result of grazing and supplemented through a combination of this with moorland burning (Watt, 1955; Burnett, 1964; Page, 1972). Bracken appears to be a unique indicator of such pressures in our island climate (Page, 1972, 1989, 1990b), while the density and extent of bracken is itself ultimately inclement to the success of most other pteridophyte species in competition with it. Much of the grazing pressure in this case is that of sheep.

Sheep have played an important role in British agriculture since the Middle Ages, when they were widely kept in lowland England. Large scale upland sheep farming probably began in the Pennines in the 12th century, spreading gradually to other areas, and reaching the Scottish Highlands in the late 18th century (Coppock, 1968, 1971). In Britain as a whole, sheep populations probably reached a peak about the 1870s; thereafter they declined somewhat in England (Hart, 1956). In Scotland, increasing populations of sheep in the 18th and 19th centuries, however, largely displaced cattle (Watson, 1932; Fraser-Darling, 1955), and agricultural observations suggest that it has been mainly over the last century that upland pastures have become heavily invaded by bracken, with sheep grazings even in the early part of this century not uncommonly becoming halved in extent in 40 years (Long & Fenton, 1938).

Similarly in Wales, areas of bracken have been estimated to have doubled in 30 years, with, in Britain as a whole, a loss rate of agricultural land to bracken of 10,360 ha per year (Taylor, 1980, 1985). Such reduction in pasture size has almost everywhere resulted in the even heavier grazing pressure of the same number of sheep (Coppock, 1964, 1971) on the progressively smaller bracken-free areas remaining – a vicious circle of events that has probably contributed increasingly in the last century, to the momentum of bracken's further spread (Page, 1982a, 1986).

Moorland burning

The intensity and extent of upland moorland burning through the last century, for the management of upland deer and grouse shootings, has certainly enhanced this rate of bracken spread, while also serving to remove other pteridophytes which are characteristically associated with old unburned heather moorland. These include especially widespread and often total losses of virtually all the clubmoss species (but notably stag's-horn clubmoss (*Lycopodium clavatum*) and fir clubmoss (*Huperzia selago*)) from areas which are burned on the most regular cycle. This subject, which has no previous literature, is under current conservational study in Edinburgh, and will be the subject of a future report.

Thus these agricultural changes of the last century, and especially those of more recent years, have, as a whole, probably had a more far-reaching effect on diminishing pteridophyte survival in these islands through a greater range of original habitats, than have most other factors added together. Unlike those considered below, the intensity of modern agriculture has further ensured that its processes have themselves added virtually no new habitats of pteridological importance to our landscape. Perhaps opportunity now exists, under the present policies of 'set-aside' and dual-use agricultural land, to begin to redress, if even on a local scale, some of the more profound of the last century's changes.

Landscapes of mineral extraction

The geological composition of the British Isles is extraordinarily diverse, with rocks which not only include some of every known geological era, but also ones of a wide variety of sedimentary, metamorphic and volcanic origins. Between them, these include rocks of acidic, neutral, basic and ultrabasic character. This diverse matrix beneath our feet has ensured an equal diversity of mineral content, and wherever any of these have been perceived as being of value to man, then attempts to extract these minerals have been made, usually resulting in much unwanted rock and low grade ore becoming dumped at the surface around its site of extraction.

With progressive metalliferous mining demise and disuse in the period of the last hundred years, many of these new and often novel habitats within the landscape have become available for plant colonisation. Although superficially inclement to plant life, pteridophytes have usually been among the chief pioneers of these newly-created landscapes, tolerating edaphic conditions seldom accepted well by flowering plant competition, with different groups of pteridophytes acting as potential indicators of the underlying toxic mineral conditions and their current state of degradation.

Landscapes of coal and shale extraction

One of the most widespread traditional mineral extraction industries in Britain has been that of coal. The coal extraction industry in Britain has had a very long history, in some areas, such as in the Forest of Dean in west Gloucestershire, stretching back to Roman times. The modern, great, grey pit heaps associated with the more mechanised aspects of coal extraction, however, which occur through many of our industrial landscapes, are largely a phenomenon subsequent to the industrial revolution, and although some may still be in a phase of active addition, many have become derelict during the last century. In addition to these, very similar shale bings, usually pink in colour, are associated with the paraffin extraction industry in such areas as the central lowlands of Scotland, and these were also accumulated mainly in the 19th century to become largely disused thereafter.

These types of landscape have formed habitats which are very different in their edaphic aspects from most natural sites. The coal pit heaps of the debris and seat-earths from around the main productive coal seams typically include thin coal fragments associated with shales, mudstones, ironstones, sandstones, grits, fireclays and sometimes thin-bedded limestones. The shale bings consist mainly of numerous, approximately coin-sized flakes of roasted shale rock, the organic component of which has largely been removed in the extraction processing. Many may contain other mineral contaminants, especially quantities of sulphur in pit heaps. For years after their deposition, some of the more organic-rich heaps emit heat through a process of slow internal combustion.

Tolerant especially of both coal and shale mine debris are usually colonising clubmosses and horsetails, which are now known from a number of widely separated sites (Page, 1988). Clubmoss records from such sites include especially stag's-horn clubmoss (*Lycopodium clavatum*), fir clubmoss (*Huperzia selago*) and occasionally alpine clubmoss (*Diphasiastrum alpinum*), and in one recently recorded site in the East Shropshire Coalfields, for example, abandoned in the mid 1960s and at an altitude of only about 370 m, all three species occur together (Box & Cossons, 1988). The presence of these species would appear to say something about age of the surface since abandonment of tipping, but being both mycotrophic and dependent largely on incoming rainwater supplies, perhaps may also be indicative of a certain improving status of air cleanliness, about which they may be valuable indicators. The horsetails are, by contrast, almost totally groundwater-dependent, and are good indicators of the presence in the tip of the simultaneous availability of a mixture of bases and available silica (the proportions needing further research, and varying with the species and perhaps ecotypes) and sometimes also of heavy metals, which are accumulated in the plants in analysable quantities. All the species are probably valuable for this purpose. Usually in the driest and most well-drained spots, even on steep slopes, common horsetail (*Equisetum arvense*) has often become the most extensive pteridophyte component of older coal-tip surfaces. The water horsetail (*E. fluviatile*) and its hybrids and the marsh horsetail (*Equisetum palustre*) grow in the edaphically similar but moister sites (Page, 1988, 1990a). On these sites, the value of these extremely long lived plants in promoting the eventual vegetation of their surfaces seems also to have been generally little appreciated.

Landscapes of slate extraction

In other regions, notably in North and south-west Wales, the English Lake District and southern Scotland, the extensive debris of former slate mines has also been added in sometimes great extent to the local landscape of these regions. All of these substrates are generally siliceous and acidic ones, sometimes with local basic enclaves, and may contain extensive mineral contaminants.

The pteridophytes which have widely colonised old slate mine tailings, as these have become progressively abandoned throughout the last hundred years and earlier, include especially: parsley fern (*Cryptogramma crispa*), hard fern (*Blechnum spicant*), woodland lady-fern (*Athyrium filix-femina*), broad buckler-fern (*Dryopteris dilatata*), yellow golden-scaled male-fern (*D. affinis* subsp. *affinis*); and, in slightly more neutral habitats, also: common male-fern (*D. filix-mas*), common golden-scaled male-fern (*D. affinis* subsp. *borreri*), and black spleenwort (*Asplenium adiantum-nigrum*). Sometimes common polypody (*Polypodium vulgare*) and, especially in western districts, western polypody (*P. interjectum*) are present, and colonies of woodland oak-fern (*Gymnocarpium dryopteris*) and beech fern (*Phegopteris connectilis*) sometimes occur in sheltered pockets where there is an accumulation of humus amongst loose rock debris. Where dampness, humidity and shade increase hard fern (*Blechnum spicant*) and woodland lady-fern (*Athyrium filix-femina*) may become very characteristic in clefts and cracks of cool rock faces around the mouths of mine-tunnel entrances to former subterranean faces, where dripping or emerging water slowly trickles from innumerable springs, the seepage paths below them often marked by stains of minerals and local growths of mineral-tolerant algae. On the shallower edges of pools below, it is not uncommon to find bright green stands of horsetails – especially water horsetail (*Equisetum fluviatile*) in the water, marsh horsetail (*E. palustre*) near the margins, and common horsetail (*E. arvense*) in the debris of the drier parts of the quarry floor.

As with coal and shale mine debris, the presence of these pteridophytes on slate mine debris would appear to say much about age of the surface since abandonment of tipping, while as with those of metalliferous mineral-extraction mines (below), the intimacy of the root proximity of these pioneer pteridophytes to the bare rock surfaces undoubtably suggests a high tolerance and likely valuable specific indicator value of different species for the mineral content of the underlying rock type, about which much research is yet needed.

Landscapes of metalliferous mineral extraction

Metalliferous minerals have been very widely extracted in Britain, especially in such regions as Cornwall (especially tin and copper), North Wales (especially copper) and the Pennines and southern uplands of Scotland (especially lead and zinc), with occasional remote mines in other areas such as at Strontian in Ardnamurchan, western Scotland, whose mines produced (and gave rise to the elemental name) strontium. The waste rock dumped around these mines almost always included appreciably remaining quantities of the metals being extracted (either as native metals or as complex ores, especially as sulphides), as well as other generally toxic mineral contaminants, typically including arsenic compounds.

Unlike the rock of slate mine debris, the rock of metal-bearing lodes often also includes many base-yielding veins, and hence where exposed, also frequently (though not always) forms habitats for those basicolous and usually calcicolous pteridophyte species which are tolerant of the mineral excesses of the individual substrate. Typical pteridophyte communities include a blend of non-rupestral and rupestral species in unusual combination. Of the normally non-rupestral species which have adapted to such sites, common male-fern (*Dryopteris filix-mas*) occurs quite widely, often accompanied by hard shield-fern (*Polystichum aculeatum*) and sometimes abundant brittle bladder-fern (*Cystopteris fragilis*). Populations of woodland lady-fern (*Athyrium filix-femina*) may be very common, and this species, plus common horsetail (*Equisetum arvense*) appear to be, in west Cornwall, among the few plants able to tolerate the very highly toxic slurry deposits generated in old tin-mine settling ponds. In southern Scotland, and perhaps elsewhere, marsh horsetail (*E. palustre*) may vigorously colonise and sometimes dominate the tailings of old lead-mine workings. Of the rupestral species which have immigrated into such sites, it is, however, the spleenworts (*Asplenium*) which dominate most old metal-mining areas in these islands. In the more basic sites, wall-rue (*Asplenium ruta-muraria*) and common maidenhair-spleenwort (*A. trichomanes* subsp. *quadrivalens*) are again the two most usual species, with green spleenwort (*A. viride* [= *A. trichomanes-ramosum* Ed.]) joining them and sometimes becoming exceptionally abundant (notably so, for example, on strontium mine tailings). In what are probably the less basic sites, the composition of the flora usually shifts towards a different trio of spleenworts, the sites sometimes becoming dominated by black spleenwort (*Asplenium adiantum-nigrum*), accompanied by more scattered delicate maidenhair-spleenwort (*A. trichomanes* subsp. *trichomanes*) and especially the highly local forked spleenwort (*A. septentrionale*). Habitats of this type are notable in the lead and copper mine sites of North Wales, where sometimes these species become so abundant that they spread to form the vegetation on adjacent drystone walls of the local rock, where they are often joined by rusty-back fern (*Ceterach officinarum*) in many areas of former metal ore mining.

There is much interest today in the colonisation of areas of industrial wasteland in Britain (e.g. Gemmel, 1977). Regrettably, there is also a tendency to group all such areas of former mining with these, and thus to fill and flatten all such pteridologically interesting sites, in the name of landscape improvement. The high specific indicator value of different pteridophyte species for the mineral content of the underlying rock type, and the research needed on this, indicated above, applies especially to those which thrive in these sites, about which very much yet remains to be learned (Page, 1978, 1979a,b). Treating each on its own merit, what could be a better argument for the careful conservation for their botanical value for future research, of some of the more pteridologically important of these old mine sites ?

Landscapes of railways and their environs

Railways have gradually come to form a myriad of new habitats which have become colonised by plants. They have somewhat inherited this position from the much earlier canals (e.g. see Busby, 1976; Page, 1988), and, like the canals and the old mine sites (above), have usually substantially added to, rather than subtracted from, the diversity of pteridological habitats in our landscape in a little more than the last hundred years.

With the invention of the steam railway locomotive in the early part of the 19th century, the great age of railways began. Stimulated by the railway's ability to transport raw materials and finished products more quickly than could the canals, the development of the railway system to most parts of Britain proceeded particularly rapidly. By 1843 there were already 2,000 miles (3,200 km) of line, and by 1845, 5,000 miles (8,000 km) had been laid (Turner, 1982).

The nature of the terrain being crossed and the desire to create only wide curves and level or only shallow trackbed gradients largely determined the types and complexity of the civil engineering constructions needed to traverse our complex terrain. Britain's rolling landscape is seldom level, and where lines entered territory such as south-west England, Wales or Scotland, and crossed the 'grain' of the landscape, innumerable civil engineering constructions were sometimes needed. These included bridges, viaducts, embankments, cuttings and tunnels. The Great Western Railway, for example, the longest-lived of the once-numerous individual railway companies in Britain, from its inception in the early 19th century to its absorption into British Railways in 1947, built no less than twelve thousand bridges and one thousand six hundred stations along about 3,600 miles (c. 5,760 km) of track, which eventually included not only main lines, but also over 150 more minor, and mostly rural, branch lines (Thomas, 1981). Today, about 65% of this system still remains in active use, with the remainder, and especially many of the rural branch lines and their stations and yards, disused since the mid 1950s or early 1960s. With track lifted, these have subsequently fallen into various states of disuse and dereliction, and where not destroyed, have offered new opportunities for plant colonisation (Sargent, 1984; Page, 1988).

In addition to the more major construction works, the railways too, like the canals, brought with them a wide range of brick and stone-built structures, as well as the ballasted trackbeds themselves. Water for locomotives was needed at stations and drainage needed in tunnels and cuttings.

Either side of the track itself, within the fenced perimeter of each line in an area that can usefully be called the 'railway corridor', exist strips of varying width, which have also become vegetated. The original construction of the railways was usually carefully engineered to re-use the materials quarried from cuttings to construct adjacent embankments. Cuttings and embankments are thus usually approximately equally numerous, and the rock substrate of embankments usually has an affinity with the native geology of the locality. Under long conditions of relatively minimal management compared with both the trackbed itself and with much of the modern countryside around them, and free from grazing by larger animals, a wide range of semi-natural vegetation has usually become established on both embankments and cuttings, with even the tops of the most exposed embankments irrigated by water from passing trains.

The landscape changes which the railways brought were thus more fundamental and far-reaching than that of any previous transport system. In their early stages of construction, considerable scars were undoubtedly created in the landscape. Gradually these softened as bare surfaces became vegetated. Over the course of 120 or more years of use, most lines, and especially the smaller rural branch lines, came to blend closely into the countryside around them. During this period, few of the original railway structures were replaced, and most of the older of those which survive today are thus seldom less than a century in age. Many of the constructions created have provided excellent habitats for a variety of plants as well as animal life and where track lengths have been abandoned, recolonisation by plant life has reached a new stage of progression.

The trackbed was always one of the most important and hence assiduously-maintained of all the physical aspects of the railway. Ballasted trackbeds are usually of hard rock such as hornfels or granite. On minor tracks and sidings, less detailed attention from weedkilling trains enabled many plants to often continue to hold their own, and with the abandonment of substantial areas of rural branchlines, the whole trackbed has often reverted at a surprising speed to a covering of plants.

The most extensive trackbed ballast pteridophyte present during the active use of many railway lines, especially in minor sidings, beneath buffers and in railway yards, has come to be common horsetail (*Equisetum arvense*). The moist interstices of the trackbed ballast itself also provides particularly suitable habitats for the success of fern gametophytes, and in high-rainfall districts, ferns have consequently proved to be amongst the early pioneers of such sites. The most frequent arrivals in these sites have been male-fern (*Dryopteris filix-mas*), broad buckler-fern (*D. dilatata*), common golden-scaled male-fern (*D. affinis* subsp. *borreri*) and woodland lady-fern (*Athyrium filix-femina*). On more exposed lengths of trackbed, such as on embankments, black spleenwort (*Asplenium adiantum-nigrum*) occurs in some western districts, whilst adder's tongue (*Ophioglossum vulgatum*) and moonwort (*Botrychium lunaria*) have occasionally also established – both the latter probably greatly benefiting from the lack of disturbance or grazing of these sites.

Embankment sides provide especially well-drained habitats, down the slopes of which the edaphic mosaic has often become diversified, and this diversity is often maintained by regular tipping of spent trackbed ballast, creating scree-like patches with a rubble-like or cindery surface. Where embankment tops are edged with stone or mortared brick retaining courses below the ballast fringe, and where shrub or tree growth is absent, ferns such as wall-rue (*Asplenium ruta-muraria*), maidenhair spleenwort (*A. trichomanes* subsp. *quadrivalens* and perhaps subsp. *trichomanes*) and black spleenwort (*A. adiantum-nigrum*) have often widely colonised. In cutting interiors, the environment becomes generally more sheltered than on embankments and edaphically more moist, and in upland areas in particular, bare rock areas, including those of basic strata, are often directly exposed. A great number of widespread ferns have come to occur in these sites, while railway-inhabiting ferns are not solely confined to species of purely local origin, and the intriguing railway distribution of maidenhair fern (*Adiantum capillus-veneris*), Font-Quer's horsetail (*Equisetum* x *font-queri*) and limestone oak-fern (*Gymnocarpium robertianum*) suggests a high value of these species as indicators of important bio-historic aspects of the railway corridor habitat in relation to more general pteridophyte dispersal (Page, 1988).

The railway corridor has thus acted as both a refuge from changing conditions around it, and as a potential corridor along which pteridophytes sometimes meet new opportunities for unusual dispersal, along features which have themselves become integral to the broader landscape of these islands.

Conclusions

The above account summarises some of the more widespread habitat changes that have occurred in our landscape over the last century, and the ways in which ferns and fern allies have responded. Overall, the greatest changes which have occurred, in terms of land area affected, have thus certainly been those of agriculture (70% of land in Britain), followed by forestry (7% of land). It is a sad irony that it is these two sources of landscape change which the above analysis shows have generally eliminated the most and added the least to the persistence of the original pteridophyte biodiversity in the last century, while much of this destruction (especially that through agriculture) has occurred with ever-increasing intensity in the post Second World War period.

The analysis also shows that other types of land use have, however, added to our landscape habitat diversity for wild species and have thereby maintained old or created new sites, albeit on a very much more localised scale, in which pteridophytes have responded, often as pioneer species, to the availability of these habitats.

In terms of the pteridophyte biodiversity which has survived the last century's revolutionary landscape changes, this account only touches on the many special aspects of their ecology which can be learned from this, through sporophyte, gametophyte and spore stages. The indicator value and general biology of the species typically encountered

(Page, 1978, 1979a,b; Dyer & Page, 1985), the dynamic processes of adaptation and natural selection influencing them (*sensu* Turesson, 1922; Turrill, 1948), the special combinations of biological strategies which have enabled each of them to succeed (*cf.* Clapham, 1956; Southwood, 1977; Grime, 1977, 1984a,b, 1986; Grime & Mowforth, 1983). The plant community as a working mechanism, with the pattern and processes of change of its populations in time (*sensu* Watt, 1947; Harper, 1982), are all fields which are, as yet, scarcely researched. These factors are of special interest within an island flora (*cf.* MacArthur & Wilson, 1967), and for some of these, aspects of field research of animal communities can provide important and sometimes stimulating analogies (e.g. Elton, 1968). For the future exploration of many of these aspects, further dedicated coupled field and laboratory studies will certainly be paramount (e.g. Steers, 1964).

Today, towards the close of the 20th century, the creation of other new habitats in our ever-changing landscape is still taking place. There are the derelict areas in many inner cities with landscapes increasingly dominated by the residue of former heavy industry. There are continuing changes in wetlands and in agricultural and forestry enterprises, influenced by changing philosophy and practice. There are the short-lived changes which occurred in bomb-sites in London during and after the Second World War, which temporarily dramatically added to inner city pteridophyte diversity in a way which none would wish to recur. There is the appearance too of other grassland areas free from major grazing which have developed this century on airfields and along our motorway network as well as from earlier origins on and around golf courses - these, plus the almost all pervasive effect of pollution, both of groundwater and atmosphere, and changes in response to potentially shifting climatic parameters (e.g. Lamb, 1985) - will have to be studies for the future.

If there are any general deductions to be made from these changes for pteridophytes in the British Isles, they are these. That the continuing process of landscape change around us steadily (and sometimes dramatically) changes pteridophyte habitats, and that while many old ones are lost, we should not ignore the fact that there is also a plus side on several occasions, where new opportunities for pteridophyte exploitation are unwittingly created by man. Many of the habitats discussed here remain still in a state of dynamic change in time, and, as pteridologists, we have to be aware of these changes, and from them very much remains to be learned. In conservation terms, we must treat each situation on its own merit, to resist, where possible the worst, but to preserve some of each of the best of these changes, and continue to observe and learn from them as integral parts of our evolving island landscape, into the 21st century.

References

BOX, J.D. & COSSONS, U. 1988. Three species of clubmoss (Lycopodiaceae) at a lowland station in Shropshire. *Watsonia* **17**: 69-71.
BURNETT, J.H. (ed.) 1964. *The vegetation of Scotland*. Edinburgh: Oliver & Boyd.
BUSBY, A.R. 1976. Ferns in canal navigations in Birmingham. *Fern Gaz.* **11**: 269.
CLAPHAM, A.R. 1953. Human factors contributing to a change in our flora: the former ecological status of certain hedgerow species. *In* Lousley, J.E. (ed.), *The changing flora of Britain*: 26-39. Oxford: Botanical Society of the British Isles.
---- 1956. Autecological studies and the 'biological flora of the British Isles'. *J. Ecol.* **44**: 1-11.
COPPOCK, J.T. 1964. *An agricultural atlas of England and Wales*. London: Faber & Faber.
---- 1968. The geography of agriculture. *J. Agric. Econ.* **19**: 153-175.
---- 1971. *An agricultural geography of Great Britain*. London: Bell.
COWAN, A. 1911. Report of the Scottish Alpine Botanical Club excursion in 1910 to the Spittal of Glenshee. *Trans. Bot. Soc. Edinb.* **24**: 171-173.
DIMBLEBY, G.W. 1978. Changes in ecosystems through forest clearance. *In* Hawles, J.G. (ed.), *Conservation and Agriculture*: 3-16. London: Duckworth.
---- 1984. Anthropogenic changes from Neolithic through Medieval times. *In* Harley, J.L. & Lewis, D.H., (eds), *The flora and vegetation of Britain. Origins and changes - the facts and their interpretation*: 57-72. London: Academic Press.

DYER, A.F. & PAGE, C.N. (eds). 1985. *Biology of pteridophytes.* Edinburgh: Royal Society.
ELTON, C.S. 1968. *The pattern of animal communities.* London: Methuen.
FRASER-DARLING, F. 1955. *West Highland survey.* Oxford: Clarendon Press.
GEMMEL, R. 1977. *Colonisation of industrial wasteland.* London: Arnold.
GIMMINGHAM, C.H. 1964. Dwarf shrub heaths. *In* Burnett, J.H. (ed.), *The vegetation of Scotland*: 232-289. Edinburgh: Oliver & Boyd.
GODWIN, H. 1975. *History of the British flora.* Cambridge: Cambridge University Press.
GRIME, J.P. 1977. Evidence for the existence of three primary strategies in plants and its relevance to ecological and evolutionary theory. *Amer. Natur.* **111**: 1169-1194.
---- 1984a. The ecology of species, families and communities of the contemporary British flora. *New Phytol.* **98**: 15-33.
---- 1984b. Factors limiting the contribution of pteridophytes to a local flora. *In* Dyer, A.F. & Page, C.N. (eds), *Biology of pteridophytes:* 403-421. Edinburgh: Royal Society.
---- 1986. The ecology of species, families and communities of the contemporary British flora. *In* Harley, J.L. & Lewis, D.H. (eds), *The flora and vegetation of Britain. Origins and changes - the facts and their interpretation*: 15-33. London: Academic Press.
---- & MOWFORTH, M.A. 1983. Variation in genome size – an ecological interpretation. *Nature* **299**: 153-155.
HARPER, J.L. 1982. After description. *In* Newman, E.I. (ed.), *The plant community as a working mechanism*: 11-25. Oxford: British Ecological Society special publication No. 1.
HART, J.F. 1956. The changing distribution of sheep in Britain. *Econ. Geogr.* **32**: 260-274.
HOOPER, M.D. 1970. The botanical importance of our hedgerows. *In* Perring, F. (ed.), *The flora of a changing Britain*: 58-62. London: Botanical Society of the British Isles.
HOSKINS, W.G. 1970. *The making of the English landscape.* London: Hodder & Stoughton.
HUNTER, R.F. 1962. Hill sheep and their pasture: a study of sheep grazing in south-east Scotland. *J. Ecol.* **50**: 651-680.
JERMY, A.C., ARNOLD, H.R., FARRELL, L., & PERRING, F.H. 1978. *Atlas of ferns of the British Isles.* London: Botanical Society of the British Isles and the British Pteridological Society.
KELLY, D.L. 1981. The native forest vegetation of Killarney, south-west Ireland: an ecological account. *J. Ecol.* **69**: 437-472.
KING, J. & NICHOLSON, I.A. 1964. Grasslands of the forest and subalpine zone. *In* Burnett, J.H. (ed.), *The vegetation of Scotland*: 168-215. Edinburgh: Oliver & Boyd.
LAMB, H.H. 1985. Climate and landscape in the British Isles. *In* Woodell, S.R.J. (ed.), *The English landscape. Past, present and future*: 148-167. Oxford: Oxford University Press.
LONG, H.C. & FENTON, E.W. 1938. The story of the bracken fern. *J. Roy. Agric. Soc.* **99**: 15-36.
MacARTHUR, R.H. & WILSON, E.D. 1967. *The theory of island biogeography.* Princeton, NJ: Princeton University Press.
MARDON, D. K. (in press). Conservation of montane willow scrub in Scotland. *Trans. Bot. Soc. Edinb.*
MITCHELL, F. 1976. *The Irish landscape.* London: Collins.
MORGAN, S.W. 1936. Domesday woodland in south-west England. *Antiquity* **10**: 306-324.
McCLINTOCK, D. 1975. *The wild flowers of Guernsey.* London: Collins.
McVEAN, D. & RATCLIFFE, D.A. 1962. *Plant communities of the Scottish Highlands.* London: HMSO.
PAGE, C.N. 1967. Sporelings of *Equisetum arvense* in the wild. *Brit. Fern Gaz.* **9**: 335-338.
---- 1972. The taxonomy and phytogeography of bracken – a review. *Bot. J. Linn. Soc.* **73**: 1-34.
---- 1978. Ferns as taxonomic tools and the future of pteridology. *Trans. Bot. Soc. Edinb.* **42**: 37-41.
---- 1979a. The diversity of ferns: an ecological perspective. *In* Dyer, A.F. (ed.), *The experimental biology of ferns*: 10-56. London: Academic Press.
---- 1979b. Experimental aspects of fern ecology. *In* Dyer, A.F. (ed.), *The experimental biology of ferns*: 551-589. London: Academic Press.
---- 1982a. The history and spread of bracken in Britain. *Proc. Roy. Soc. Edinb.* **81B**: 3-10.
---- 1982b. *The ferns of Britain and Ireland.* Cambridge: Cambridge University Press.
---- 1986. The strategies of bracken as a permanent ecological opportunist. *In* Smith, R.T. & Taylor, J.A. (eds), *Bracken. Ecology, land use and control technology*: 173-181. Carnforth: Parthenon Publishing.
---- 1988. *Ferns. Their habitats in the British and Irish landscape.* London: Collins New Naturalist.
---- 1989. Three species of bracken, *Pteridium aquilinum* (L.) Kuhn, in Britain. *Watsonia* **17**: 429-434.
---- 1990a. Hybrids in the genus *Equisetum* in Europe: an updated annotation. *In* Rita, J. (ed.), *Taxonomia, Biogeografia y Conservacion ed Pteridofitos*: 1-6. Palma de Mallorca: Universitat de les Illes Balears.

---- 1990b. Taxonomic evaluation of the fern genus *Pteridium*, and its active evolutionary state. *In* Thomson, J.A. & Smith, R.T. (eds), *Bracken biology and management*: 23-34. Australian Institute of Agricultural Science, Occasional Publication No. 40.
PATTON, D. & STEWART, E.J.A. 1914. The flora of the Culbin Sands. *Trans. Bot. Soc. Edinb.* **26**: 345-374.
---- ---- 1924. Additional notes on the flora of the Culbin Sands. *Trans. Bot. Soc. Edinb.* **29**: 27-42.
PEARSALL, W.H. 1950. *Mountains and moorlands*. London: Collins New Naturalist.
PERRING, F. 1974. *The flora of a changing Britain*. London: Botanical Society of the British Isles.
PETERKEN, G.F. 1974. A method for assessing woodland flora for conservation using indicator species. *Biological conservation* **6**: 239-285.
---- 1981. *Woodland conservation and management*. London: Chapman & Hall.
POORE, M.E.D. 1985. Agriculture, forestry and the future landscape. *In* Woodell, S.R.J. (ed.), *The English landscape. Past, present and future*: 188-201. Oxford: Oxford University Press.
RACKHAM, O. 1976. *Trees and woodlands in the British landscape*. London: Dent.
---- 1980. *Ancient woodland*. London: Arnold.
---- 1986. *A history of the countryside*. London: Dent.
RATCLIFFE, D.A. 1968. An ecological account of the Atlantic bryophytes in the British Isles. *New Phytol.* **67**: 365-439.
---- 1977. *Highland flora*. Inverness: HIDB.
---- 1984. Post-Medieval and recent changes in British vegetation: the culmination of human influence. *In* Harley, J.L. & Lewis, D.H., (eds), *The flora and vegetation of Britain. Origins and changes – the facts and their interpretation*: 73-100. London: Academic Press.
RAVEN, J. & WALTERS, M. 1956. *Mountain flowers*. London: Collins New Naturalist.
SARGENT, C. 1984. *Britain's railway vegetation*. Huntingdon: NERC.
SOUTHWOOD, T.R.E. 1977. Habitat, the template for ecological strategies? *J. Animal Ecol.* **46**: 337-365.
STAMP, L.D. 1955 (4th ed.). *Britain's structure and scenery*. London: Collins New Naturalist.
STEERS, J.A. 1964. *Field studies in the British Isles*. London: Nelson.
STEWART, O. 1988. *Pilularia globulifera* in Kirkudbrightshire. *Trans. Dumfries. & Galloway Nat. Hist. & Antiq. Soc.* ser. 3, **63**: 1-4.
SYDES C. & MILLER, G.R. 1988. Range management and nature conservation in the British uplands. *In* Usher, M.B. & Thompson, D.B.A. (eds), *Ecological change in the Uplands*: 323-337. Oxford: Blackwell Scientific Press.
TANSLEY, A.G. 1939. *The British Islands and their vegetation*. Cambridge: Cambridge University Press.
---- 1949. *Britain's green mantle*. London: Allen & Unwin.
TAYLOR, D.B. (ed.), 1979. *The third statistical account of Scotland. Vol. 27. The counties of Perth and Kinross*. Couper Angus: Culross the Printers.
TAYLOR, J.A. 1980. Bracken: an increasing problem and a threat to health. *Outl. Agric.* **10**: 298-304.
---- 1985. The relationship between land-use change and variations in bracken encroachment rates in Britain. *In* Smith, R.T. (ed.), *The biogeographical impact of land use changes*: 19-28. Norwich: BSB/Geo Books.
THOMAS, A.S. 1963. Changes in vegetation since the advent of myxomatosis. *J. Ecol.* **48**: 287-356; **51**: 151-186.
THOMAS, V.G. & GEORGE, J.C. 1975. Plasma and depot fatty acids in Canada geese in relation to diet, migration, and reproduction. *Physiol. Zool.* **48**: 157-167.
---- & PREVETT, J.P. 1982. The role of horsetails (Equisetaceae) in the nutrition of northern-breeding geese. *Oecologia* **53**: 359-363.
THOMAS, D.S.-J. 1981 (5th ed.). *A regional history of the railways of Great Britain. Vol. 1. The West Country*. Newton Abbot: David & Charles.
TURESSON, G. 1922. The genotypical response of the plant species to habitat. *Hereditas* **3**: 21-250.
TURNER, D. 1982 *Railways in the British Isles. Landscape, land use and society*. London: A. & C. Black.
TURRILL, W.B. 1948. *British plant life*. London: Collins New Naturalist.
WALTERS, M. 1984. The relation between the British and European floras. *In* Harley, J.L. & Lewis, D.H. (eds), *The flora and vegetation of Britain. Origins and changes – the facts and their interpretation*: 3-13. London: Academic Press.
WATSON, J.A.S. 1932. The size and development of the sheep industry in the Highlands and north of Scotland. *Trans. Highland Agric. & Statistical Soc.* ser. 5. **44**: 1-25.
WATT, A.S. 1947. Pattern and process in the plant community. *J. Ecol.* **35**: 1-22.
---- 1955. Bracken versus heather, a study in plant sociology. *J. Ecol.* **43**: 490-506.
WEBSTER, M.M. 1978. *The flora of Moray, Nairn and East Inverness*. Aberdeen: Aberdeen University Press.

WEBB, D.A. 1983. The flora of Ireland in its European context. *J. Life Sci. Roy. Dublin Soc.* **4**: 143-160.

WHITE, D.J.B. 1961. Some observations on the vegetation of Blakeney Point, Norfolk, following the disappearance of the rabbits in 1954. *J. Ecol.* **49**: 113-118.

Note: the English names for pteridophytes used by Page are not widely accepted. They have been left unamended to enable the reader to correlate this text with his books *The Ferns of Britain and Ireland* (1982) and *Ferns. Their habitats in the British and Irish Landscapes* (1988). Ed.

Chris Page omits only molecular science from the spectrum of his interest in gymnosperms (especially conifers) and pteridophytes (especially Equisetum*). An author and occasional broadcaster who has travelled widely in the British Isles, Australasia, the eastern Pacific and North America, he enjoys applying his knowledge to conservation, ecological changes and the development of arboreta in the British Isles.*

My Life with Ferns
R. E. Holttum†

I was born at Linton, on the chalk in the south of Cambridgeshire, in 1895. I early became interested in the plants of the countryside, but there were no ferns, not even bracken. Later I went to the Lake District, but came in contact with no-one specially interested in ferns. At the University of Cambridge the professor of botany (A. C. Seward) was primarily interested in fossil plants, so I was taught about the biology and anatomy of ferns with especial emphasis on those thought to be primitive and thus related to fossils.

In 1922 I was appointed Assistant Director in the Gardens Department, Straits Settlements and stationed at Singapore. I was asked by the Director (I. H. Burkill) to take a special interest in ferns. The area open for field work was the Malay Peninsula, and in the herbarium were many specimens from Indonesia and the Philippines, obtained by exchange. There were about five hundred native species in the Peninsula (as compared with about fifty in Britain, of about the same area) and far more in the Malayan region as a whole. I found that there were three books which would help me to identify the ferns so abundantly around me, but they had three different schemes of classification. The only recent one was by van Alderwerelt van Rosenburgh (based at Bogor in Java) who was not interested in genetical (or supposed genetical) relationships and devised a new scheme of classification of his own based on what seemed to him convenient characters.

Fortunately, at an early stage of my studies in Singapore, I began to correspond with Carl Christensen of Copenhagen. His first great work had been to compile an *Index Filicum* including all published fern names. This led him to think about the lack of clear generic concepts shown in past literature, and his first subsequent work was to study the tropical American ferns which had been considered related to the well-known British species *Dryopteris filix-mas*. By looking at characters not previously noticed by fern taxonomists, he produced for the first time clear generic distinctions within the group of species with which he was concerned. In the 1920s he was becoming interested in the ferns of Asia and Malesia, and his guidance and his identification of my specimens helped me greatly to come to a better understanding of a very confused and complex subject.

In 1926-27 I had leave of absence from Singapore to Britain, and then came into contact with British ferns with a better understanding; I thought they were a strange-looking lot of odds and ends which appeared to need more study to connect them with the increasing studies (not only Christensen's) of the world's ferns as a whole. But I did become a member of the British Pteridological Society. The best-known British Flora at that time was J. D. Hooker's revision of Bentham's *Handbook to the British Flora* (originally published 1858). J. D. Hooker still adopted his father's classification in his great work on the world's ferns, *Species Filicum* (five volumes, 1846-64). For example, the beech fern was named *Polypodium phegopteris* because its sori lack indusia, though it is now evident that it is not at all nearly related to the type species of *Polypodium*, *P. vulgare* [= *P. cambricum* Ed.]. I had become acquainted with many species of the *Polypodium* family, mostly epiphytes, in Malaya.

The next stage of my fern-study was due to contact with R. C. Ching who called on me at Singapore in 1929 when he was on his way to Europe to study ferns with Christensen. I corresponded frequently with him until 1941 and learned much also from his publications on the ferns of mainland Asia, to which Malesian species are allied. During the Japanese occupation of Singapore (1942-1945) I wrote a full systematic account of the ferns of the Malay Peninsula, so far as then known to me. This was

†died 18 September 1990. Obituaries in the *Fern Gazette* **13**: 383; *Bulletin* **4**: 30-31.

later revised, and published in 1954, at the time of my retirement from Singapore. Thereafter I lived at Kew, working in the Herbarium of the Royal Botanic Gardens as an honorary research associate, making detailed monographic studies of a succession of groups of Malesian ferns for *Flora Malesiana*. I also made many travels to see specimens of ferns in other great herbaria.

Meanwhile, various authors had published different schemes of classification of ferns into families. In my book of 1954 I had adopted a scheme of my own which placed many genera of Malayan ferns into a family Dennstaedtiaceae. But when I came to write an introduction to ferns in *Flora Malesiana* (1959) I did not feel able to adopt a definite family allocation for some genera; I arranged them in groups, retaining family names only where I believed that there was very clear evidence.

My monographic studies for *Flora Malesiana* were: Gleicheniaceae and Schizaeaceae (1959), Cyatheaceae (1963), the Lomariopsis group of genera (1978) and Thelypteridaceae (1981). An account of *Tectaria* and allied genera is complete but not yet published. The Lomariopsis group of genera was one to which I had devoted much field study and had for the first time described their peculiar growth-habits. The only one of my Malayan monographs concerned with a family represented in Europe was that on the *Thelypteris* family. This is a quite distinct family, though not recognized as such until a paper by Ching in 1940, and it is one of the largest fern families. In the 19th century members of this family were confused with *Dryopteris* and other genera because no-one had noticed its distinctive characters (Christensen was the first to do so).

In the British Flora now most commonly used (by Clapham, Tutin and Warburg, first published in 1952) all leptosporangiate ferns are included in one family named Polypodiaceae, divided into seven subfamilies but without reference to ferns in them in other parts of the world. There can be little doubt that leptosporangiate ferns constitute a natural group, but their great diversity (and great number of species) in the whole world make a division into families appropriate, though there is still no agreement as to the delimitation of some of them and even less agreement as to their inter-relationships. I cannot attempt a critical statement on all C. T. & W.'s subfamilies and will conclude this paper by commenting on the family Thelypteridaceae, which comprises nearly one thousand species in the whole world.

For Flora Malesiana I spent about twelve years of intensive study on this family, describing 440 species in 22 genera of which seven were newly recognized and described by me in the course of this work. The area covered by Flora Malesiana consists of Malaysia, Indonesia, the Philippines and the whole of New Guinea. I also examined specimens of all species recorded as occurring in mainland Asia and the Pacific, and incidentally looked at those of Africa where species related to Asia and America co-exist.

A combination of two characters is distinctive of the family: the arrangement of vascular strands in the stipe (two at the base, uniting upwards to form one which is U-shaped in section) and the presence of unicellular, acicular hairs on the adaxial (upper) surface of rachis and costae. Fronds are almost always simply pinnate with many pairs of pinnae almost equal in length, each pinna symmetrical at its base on each side of the costa, the pinnae lobed with a translucent membrane in the base of each sinus between lobes. Chromosome numbers in the family are 27-36, thus differing from the 41 present in *Dryopteris*. Three British species belong to the family. In C. T. & W. they are named *Thelypteris palustris, T. limbosperma* and *T. phegopteris*, but on a world scale they are regarded as belonging to three genera. C. T. & W. also include in *Thelypteris, T. dryopteris* (L.) Slosson and *T. robertiana* (Hoffm.) Slosson, but these do not belong to Thelypteridaceae as above defined; they represent a genus *Gymnocarpium* Newm. more nearly related to *Dryopteris*.

Thelypteris palustris Schott is the type species of its genus. As now restricted the genus consists of one widely distributed northern species which includes sub-species, and one southern species in Africa, Madagascar, South India, Thailand, Sumatra, New

Guinea and New Zealand, in Malesia on mountains in open swamps near lakes.

The genus *Phegopteris* consists of three species. the British one was first named *Polypodium phegopteris* by Linnaeus and is now known as *Pheopteris connectilis* (Michx.) Watt, widely distributed in north termperate regions. The other two species are distributed in eastern N. America and in S. E. Asia and Malesia; in Malesia the latter species is only known to occur in Sulawesi and East Java.

The genus *Oreopteris* consists of three species of northern temperate distribution. The British species, *O. limbosperma* (All.) Holub, is widely distributed and very variable, as recently described by Martin Rickard in the *Pteridologist* vol. 1 part 6 (1989) pages 272-273.

In Spain are two more species of the family; both belong to mainly tropical genera in which the veins anastomose in a way peculiar to the family. The genera are *Stegnogramma* and *Christella*. *Stegnogramma* comprises about 18 species. *S. pozoi* (Lagasca) K. Iwats. is variable, very widely distributed in Africa, with a distinct variety in Ceylon and Java; the type of the species was collected in Spain.

Christella is a genus of about fifty species in the palaeotropics, with 15-20 more in Africa and the neotropics which are not yet clearly defined. The Spanish species is *C. dentata* (Forsk.) Brownsey & Jermy, a tetraploid which is very widely distributed in the palaeotropics and subtropics; in Malaya it grows in lightly shaded places, not in primitive forest, and is almost a weed.

I wrote a detailed account of Thelypteridaceae in Europe, published in *Acta Botanica Malacitana*, vol. 8 (1983) pages 47-58.

CYATHEA DEALBATA. DICKSONIA ANTARCTICA. DICKSONIA SQUARROSA.

An Amateur's Steps in Pteridology
C. R. Fraser-Jenkins
71 Abingdon Road, Oxford, Oxon OX1 4PR

My fascination with ferns began to develop when I was a youngster of ten years old. I had been brought up with a keen interest in all aspects of nature, whether wild flowers, which my father would tell me the names of on country walks near our home in Bridgend, or butterflies and beetles, or crabs and lizards down by the beach. I was always enthralled by the precise and constant distinctness of each of the multitude of different types of plants and animals, their recognisability, which few other children seemed to be able to see or be interested in. At my prep. school, surrounded by the woods and slopes of the Malvern Hills, I remember one summer evening after 'lights out' escaping through the dormitory window in my pyjamas and climbing down the roof to the path leading up to the ridge of the hills just above the school – and as a souvenir, when I crept back in, I brought with me a large fern-frond from beside the path that had taken my fancy. The next day I used my pocket-money through the school's shopping system to buy *An Observer's Book of Ferns* and I was delighted to be able to recognise that my special frond could only be a fern called the lady fern, *Athyrium filix-femina*, pictured there just like the one I had before me. I repeated its mouthful of a name until it rolled easily off the unaccustomed tongue and decided, as much as one ever actually decides things, that I would collect ferns from then on. The earliest specimen I kept was from a family holiday in Jersey that summer, which I had discovered from my book was the rare *Asplenium billotii* [= *A. obovatum* ssp. *lanceolatum* Ed.], much to my surprise. This later became CRFJ no. 1 in the collection of ferns I built up, stuck onto large card sheets, at my public-school, and is now preserved in the herbarium of The Natural History Museum, London.

It became my hobby then to collect fronds of all the different ferns I could, rather like a stamp collection, and I am glad, looking back on it, that my interest was encouraged by a most enlightened house-master at Radley who helped remove some of the pressure to conform to the games ethic and similar activities that were the uninspiring norm both then and, probably, now. I was the only person allowed to put 'fern-collecting' as my obligatory form of exercise, as I cycled eagerly all around the Abingdon and Oxford area, including visiting the fascinating fern-collections at the Oxford Botanic Gardens, looking at and collecting ferns and their allies and from time to time persuading long-suffering but sympathetic members of the staff to take me somewhere to see some particular one as my catalogue slowly increased its scope and ferns increased their grip!

It was my great good luck that from the famous Professor Warburg, whom I met at Oxford when I went to visit the herbarium, I came to hear about a veritable oracle of fernery living on a farm not too far away, Hugh Corley, near Faringdon. So I wrote to him and was soon cycling over there on the first weekend I got permission to go off. He introduced me to his marvellous and original world of the ferns that were not, at that time, in the books, particularly to the specialist taxa within the male fern group (in a wide sense), which he went collecting and studying, bare-foot, in the west Scottish Highlands every summer and so had all the different forms growing in his garden. Catching part of his vast and critical enthusiasm, I plunged delightedly into this melting pot of taxonomy and under his influence began at the age of fifteen to start my own taxonomic study of ferns – a stage beyond the mere 'stamp-collection' phase, exciting and formative though that had been and long continued to be. As I visited him more often he introduced me to the study of detailed characteristics such as spore-size measurement (which he remains an unsung master of, with many remarkable new and unpublished findings), pinnule-toothing, indusial features, scales, and, above all, frond texture, showing me new and exciting finds of hybrids and proposed new genomic combinations, whose coding he pioneered, within *Dryopteris affinis* among others. I soon became familiar with the various tricky complexes of British ferns, at that time in the hey-day of being elucidated by the Leeds school of cytologists led by Professor Irene Manton. My interest was rapidly becoming both detailed and all-enveloping.

When I left school and entered university, having chosen Leicester, where the great taxonomist, Professor T. G. Tutin, could take me under his wing, there is no doubt that my one-track interest in fern taxonomy made me push aside all the rest of my courses. Though I have never had to have the slightest regrets about this, as a confirmed pteridologist, it was only through Prof. Tutin's tolerance and insistence behind the scenes that I was allowed no less than three resits of my first-year exams, much against the rules, and could scrape through my degree in 1969 and commence a Ph.D. research programme, though I later had to cut that short while teaching for a few years. Indeed even while applying for various universities the problem had already shown itself in my non-botanical academic work in the form of my only having two 'A'-levels, which earned the fair comment from Professor Manton, who interviewed me for Leeds, that I hadn't a hope of getting into any university, though she herself kindly swung things my way at Leeds both then and later. This had all been foreseen by another early mentor of mine, the spirited Dr Hugh Cardwell at Radley, who had wisely aimed me for Leicester after helping to secure an offer of a place at Oxford as he knew I would not have got through the first year's more general exams otherwise. Pteridology had become too powerful!

Shortly before I left school, and continuing during university years and ever since, for that matter, I started my first botanical expeditions abroad, making extensive herbarium collections of European ferns and rapidly expanding my horizons and the number of fern species I knew. As a result it was possible to gain a much better idea than from merely British studies of both geographical variation within species and the markedly different 'uniqueness' of any genuinely distinct species, as opposed to mere variants. I had informally connected myself to the British Museum (Natural History) at that stage, partly through my previously having joined the British Pteridological Society as its youngest member and lapping up the fascinating meetings at the Museum, and partly with the great help and encouragement of Mr Clive Jermy there, who fostered my herbarium collecting with both instruction and Museum grants. Most importantly he also introduced me to proper taxonomic method and literature by obtaining a formative Vacation Research Studentship for me at the BM, where, needless to say, I indulged to my heart's content in the preliminary study of the *Dryopteris villarii* and *D. pallida* aggregate, which I later published in 1977 and 1980. In 1968 I joined Dr Martyn Rix on a Cambridge University expedition to Turkey and Iran and, following Hugh Corley's male-fern training, I was lucky enough to discover what I was immediately suspicious was the missing diploid ancestor of *Dryopteris filix-mas*, Hugh's '*incognita*'. Continuing to expand my botanical trips in 1969 and 1970, I went back to the Pontic fern-forests and also to the Soviet Caucasus, at that time little known to western botanists, and collected more of this 'unknown' species, becoming convinced that it was indeed what we had been looking for. Hugh Corley, too, was certain of it, and the breakthrough when I first found it indeed to be diploid was one of the great excitements of my pteridological life. It was at the instigation of Professor Manton that I had been invited to join in for a term at Leeds University in Professor John Lovis' systematics course, which included instruction in carrying out chromosome squashes and the detailed study of the *Dryopteris filix-mas* group from Brathay in the Lake District which we used as sporing and cytological material. As one of only three students, another being Dr Mary Gibby as an undergraduate, we were able to benefit enormously from the friendly and detailed supervision of the then Dr Lovis, who was the leading expert on *Asplenium* cyto-systematics and a most active fern-worker. Having just returned from the Caucasus bearing my 'missing ancestor' *Dryopteris*, I was delighted that we could include it in the course, which was how I could investigate it cytologically, and which a cautious, but finally convinced Dr Anne Sleep confirmed later that evening. This discovery of what was to become *Dryopteris caucasica* in my 1972 paper with Hugh Corley was a most welcome crowning of his ideas, as also those of Professor Manton herself, who was equally excited by it immediately on seeing the fronds; she too could see it was a pretty sure candidate for the missing parent even before it was confirmed cytologically.

A few months later, as a result of this, I received out of the blue a most interesting letter from Professor Tadeus Reichstein of the University of Basel, Switzerland, asking for material for chemical analysis. His enquiry began a long association between us in which we collaborated in the study of the remaining European *Dryopteris* complexes in a series of joint and independent publications. It has also been my great good fortune that his long-term financial help has allowed me to continue my interest and work as an amateur pteridologist, a field with few professional posts available, now threatened with extinction in Britain, and in which I am in the almost unique position of keeping myself going independently, so far, though it is not always easy to manage it. But I have constantly been able to travel all over the world looking at *Dryopteris*, and the related genera *Polystichum* and *Athyrium*, as a result of his requests to collect obscure *Asplenium* species for his hybridization programmes and with his backing. It was with his help that I was able to expand beyond the European fern flora (in the wide sense), which I felt I had come to grips with in 'my' groups, to the Himalaya, the next area eastwards, where my sights began to set in 1977 when I first crossed the desert and drove over to India by road.

Here I was confronted suddenly with an astonishing diversity of genera and species, and I set out hoping to begin the treatment of *Dryopteris* in the same sort of cyto-systematic way as the European ones. But instead of the mere fragment of the genus in Europe (only 10 British and 21 European ones) there were now no less than 57 Indian subcontinent species, not to mention their varieties (a figure that emerged later from my study), none of which I knew at all as I had previously been putting them aside in my herbarium studies as 'something tropical'. In addition they represented a far wider morphological cross-section of the genus than the few random groups represented in Europe. Within a short time I realised that to understand them I would have to learn the Chinese, Taiwanese and Japanese species as well, and since south-west China is the centre of richness of the genus and China alone contains about 110 species, I had my hands full – and worse, not one of the polyploid complexes had been 'worked out' or understood by means of genome analysis even though there are multitudes of tetraploid sexual, and diploid or triploid apomictic species, all derived originally by hybridization, plus several additional sterile hybrids! It was obviously necessary for the time being to return to traditional taxonomy, though I was helped greatly by the cytological study carried out on my living collections by Dr Mary Gibby, by now a top-rank cytologist at the British Museum (Natural History). I was also lucky enough, with the help of Clive Jermy, to obtain a major three-year Nuffield Foundation Fellowship to work at the BM on Himalayan *Dryopteris*, *Polystichum* and *Athyrium*, the three largest genera apart from *Asplenium*, and two Royal Society Scientific Research Fellowships to enable me to exchange with the Academy of China and work there on the relevant Chinese species. The results have been that some 12 years further on I feel that I have been able to get to grips with Indian *Dryopteris*, published in a detailed monograph, and with *Polystichum* and *Athyrium*, also either published or about to be. Complimentary phytochemical papers on Asian *Dryopteris* are also under way as a result of joining forces with Professor C.-J. Widén of Helsinki, to obtain further 'hard evidence' to separate taxa. With Professor Reichstein's help I have also been able to visit Africa, Central America and Hawaii – all areas where the genus is somewhat complicated – and have now obtained a fully world-wide detailed view of the genus which was published as a summarised monographic classification four years ago, though there are some further local modifications necessary to it. Before long I have in mind to do the same sort of study with *Polystichum*, which is under way at present.

In the meantime the 'stamp-collecting' aspects of pteridology, in the way of cataloguing, have metamorphosed into a more particular interest in the ferns of the Sino-Himalayan region, the simpler, western-most parts of which I am now quite at home in after frequent sojourns there, resulting in accounts of the ferns for *Flora Iranica* (including western Pakistan) and the *Flora of Pakistan*, both soon to be published – though for me the real excitement lies over in the east Himalaya where the ferns are so diverse and numerous that it takes years to get to know them, only to find there are always more than one had thought. But curiously, the longest-term and most detailed study I have undertaken so far, which will hopefully soon be complete, is one that I have been involved with all through my pteridological career involving the European species first shown to me by Hugh Corley in about 1963, *Dryopteris affinis*. It is by far the most complicated species I have ever come across and has required by far the most critical and careful study, involving geographical variants, different genomically-based (as far as I can understand them), apomictic subspecies, several different sterile hybrids and all the pitfalls of variation and semi-cryptic taxa. It has given me a sense of admiration to discover after all these years of work that Hugh Corley's original ideas on the group have all been borne out on investigation, covering the major part of the group.

Despite the gradual expansion of detailed taxonomic knowledge I have been going through, I find that I have ended up coming back to base in some respects – it is only over the last four years that I have begun to discover what others often began their interest with, something I too once knew, but tended to lose on the way, as a result of too much naming and cataloguing to have taken the time off to appreciate it. While it is possible, I am lucky enough now to be able to return to a more full enjoyment of the unfathomable perfection and 'is'-ness of all the different ferns by growing them in my father's garden in south Wales, back to where it all began, along with seeing ferns in the collections of other enthusiastic Pteridological Society members all over Britain, particularly Martin Rickard into whose fine garden I have introduced more than a few exotic species. In many ways this has given me a fuller satisfaction than the rather hard-driven scientific quest of taxonomy itself, even if that still continues apace.

A Brief History of Ferns and their Cultivation

Matthew V. Ford

Royal Botanic Gardens, Kew, Richmond, Surrey TW9 3AB

Ferns have had quite a chequered history, being the subject of great myths, and then such mania that is now reserved for the likes of the World Cup. The following gives an outline of Man's involvement with ferns, from the earliest times to the era of test-tube plants.

The fern has been the source of folklore and mysticism and has been an inspiration for poets and painters. Britten (1881) is one of several authors who give many details of folklore and uses of ferns. Only a few examples are given here. The moonwort (*Botrychium lunaria*), being associated with things magical for a long time, was held in great respect. During mediaeval times it was believed to have the power to open locks, unshoe horses and change metal. Fairies used moonwort to saddle their horses. Shakespeare tells us that Queen Mab saddled her steed with the pinnules of moonwort and galloped *'night by night through lovers' brain, and then they dreamed of love'* (Heath, 1898). The male fern, *Dryopteris filix-mas*, was seen as essential in a love potion, and would be more successful if mixed by a witch. Drying the rhizome in the smoke of a midsummer fire and fashioning it into the shape of the fingers of a hand made a great good-luck charm.

The all powerful Doctrine of Signatures was a creed concerned with, amongst other things, the interpretation of patterns and shapes among the connecting bundles when cutting roots, stems and leaves of plants horizontally. This greatly influenced the superstitious of olden days. In the brake (*Pteridium aquilinum*) the shape 'C' was found and seen as the initial of Christ, thus prized as protection against witches and goblins. In Scotland an 'x' was usually discerned, again a symbol for Christ, but in some areas a shape seen as the devil's hoof was found!

The wonder and speculation over how ferns reproduced, which was debated for centuries, led people to believe that the 'invisible' seed had the power to impart its invisibility to the finder. In Shakespeare's King Henry IV, Chamberlain says to Gadshili *'You are more beholding to the night than to fern seed for your walking invisible'*. Also from the play Gadshill exclaims *'We steal as in a castle, cock-sure: we have the receipt of fern seed, we walk invisible'*.

Many pagan ceremonies sprang up around this belief. One involved taking twelve pewter plates on midsummer's eve and placing them under the *'black spotted frond'*. The magical seed would pass through eleven of the plates and rest on the twelfth. Sometimes, however, fairies snatched the precious seed away as it fell, watched over by Oberon, the fairy king. If the gatherer was successful then they would be endowed with the much desired quality of invisibility and rewarded with the protection of spirits and realisation of all their fondest desires. This superstition persisted among the country people of Worcestershire until as late as the nineteenth century. (From Heath, 1898.)

Equally practical, and more believable today, is the traditional use of ferns for medicinal purposes (May, 1978; Abbe, 1981). As far back as Dioscorides and Galen of the first and second centuries respectively, they have been accepted as good drug sources. So far ferns have been largely ignored in the recent interest in pharmacognosy, but this aspect is receiving attention in China (Luo, 1989; Wang, 1989).

Folklore went as far as to describe the flower and timing of seed dispersal. Some had the flower as a wonderful globe of sapphire blue and some as ruby red but only blooming at midnight. Moments after flowering, the seed appeared.

In the early 16th century Pliny stated that the fern had neither flower nor seed and some of the old English writers believed in this. Britten (1881) recounts William Turner's mid-sixteenth century translation of how Tragus, a mediaeval *'phisicion'*, set out to investigate this. *'. . .I have foure yeres together one after an other, upon the vigill of saynt John the Baptiste (whiche we call in Englishe mydsomer even) soughte for this sede of Brakes upon the nyghte, and indede, I founde it earlye in the mornynge before the daye brake; the sede was small blacke, and lyke unto poppye. I went aboute this busynes all figures, conjurynges, saunters, charmes, wytchcrafte, and sorseryes sett a syde, takynge wyth me two or three honest men ... Sometyme when I soughte the sede I founde it and sometyme I founde it not. Sometyme I founde muche and sometyme lytle; but what shoulde be the cause of this diversyte or what nature meaneth in this thinge, surelye I can not tel.'*

In early times, there were very few fern collections in England. In 1628 John Tradescant, who owned a botanic garden and museum at Lambeth, brought some rare plants back from Virginia, including *Cystopteris bulbifera* and *Adiantum pedatum*. Later, in 1680, he added *Camptosorus rhizophyllum*, and, in 1699, *Onoclea sensibilis*, *Adiantum reniforme* and *Davallia canariensis*. The only other recorded exotic fern in England at this time was the *Blechnum australe* in King Charles II's garden in 1671.

By the 1770s collecting ferns began to gain popularity and be considered important botanically. Between 1770 and 1790, 68 species were brought into Britain, most going to private collections, the rest to the Royal Botanic Gardens at Kew. Joseph Banks, botanical advisor to George III, made efforts to accelerate the introduction of new and rare plants, the most significant being the persuasion of the captains of ships to take an interest. In 1768 (Hill, 1769) there were 10 ferns listed in Hortus Kewensis, by 1789, 34 (Aiton, 1789). When a plant died it was replaced by another living specimen, imported from abroad. Shipping space being at a premium, priority inevitably went to the plants most in favour with the wealthy, reliable horticulturists. Kew's collection was boosted in 1795 with the return of Rear Admiral Bligh (famed for his role in the mutiny on the Bounty) from the Indian colonies. From this voyage 37 ferns, and many other plants, were contributed. (Smith, 1866, provides this historical information.) Only a year before, a report from John Lindsay to Joseph Banks was made on the propagation of ferns from spores. Lindsay, a surgeon in Jamaica, had been doing experiments on this since 1786. Banks was impressed and read a paper by Lindsay (Lindsay, 1794) on the subject to the Linnean Society. There had, actually, been a few successful attempts previously. In 1699 Robert Morison reported that ripe spores of *Osmunda regalis* and *Phyllitis scolopendrium*, shed onto moistened soil, produced '*plantulas*', which bore a succession of different types of leaves. From his reports these were the gametophyte and sporophyte stages. In 1788 Friedrich Ehrhart made records of his observations of germination of spores in several hardy native fern species. These were not followed up, despite the fact that they must have realised the horticultural implications. Lindsay was thus given the credit for developing a reliable method of propagation.

By the early nineteenth century Lindsay's method was widely used. However, some years later, when John Shepherd and his nephew Henry (successive curators of Liverpool Botanic Gardens) began to receive acclaim for their fern raising skill, the impression given in the literature (Smith, 1820) was that the method they employed had been developed independently. They were aware, though, of the successes that had been achieved before them.

Even with publicity, interest was slow to spread. Conrad Loddiges and Sons (Hackney Nurserymen) took up the torch having been instructed by and supplied with spores by the Shepherds. In 1866, John Smith, Curator of the gardens at Kew, credited the spread of fern cultivation to Messrs Loddiges. From 1823 Kew's collection was largely built up from this source, though many were grown from spores on herbarium specimens, having previously relied mainly on spore collection from the wild.

The nineteenth century saw a steady stream of new arrivals to Kew from new colonies and foreign places. This began in 1808 with *Platycerium bifurcatum*, *Doodia aspera* and *Davallia pyxidata* from Australia and, in 1816, several more species were brought from Australia, and the first from New Zealand. The arrivals were well publicised and attracted large crowds. All this time the ever growing collection of ferns at Kew had been under the care of John Smith, curator of the gardens and responsible for the major acquisitions. When he arrived in 1822, he found 40 species of ferns in the gardens, with only a couple from the tropics. Bad construction of the brick flues in the hot houses gave an atmosphere that was much too dry and full of smoke and fumes, effectively prohibiting the growth of tropical fern species. He put them in order and organised the collection of new plants, making Kew the centre of cultivation in Great Britain. By 1845 a five-fold increase had been achieved, and by 1846 there were 348 species in the fern houses (Smith, 1846).

From then on every opportunity was exploited to increase the numbers in public and private collections. Contributions came in from all over the World. A New Zealander, Mr Edgerly, caused a sensation in 1841 when he supplied the first tree-ferns, *Dicksonia squarrosa* and *Cyathea medullaris*. These were greeted with delight and were soon in great demand. Williams (1868) wrote of the International Exhibition in 1866 '... *indeed, but for the tree ferns, the Exhibition would have lacked half its beauty and attraction; no other plants we have in cultivation could have been substituted*'.

It was not only through the collections at Kew that ferns grew in popularity. Publications reached a very wide audience. From 1840-1865 Sir William Hooker was director of the gardens and published volumes of the *Species Filicum* (1846-1864), describing all known species of typical ferns. The *Synopsis Filicum* was published after his death by Mr Baker FRS, Keeper of the Herbarium (Hooker & Baker, 1868). John Smith published two books, many official listings and generally promoted the Kew collections (see Holttum, 1967). He maintained a friendly rivalry with other collectors and writers as they did with him. It was, after all, a gentlemanly pursuit. By 1857 his catalogue of cultivated ferns listed 560 exotic species in British gardens. By 1895 Kew's collection had built up to 1,116 species and varieties of exotic ferns, 97 fern allies, and 586 British ferns. Smith (1895) wrote '*The collection of ferns, whether tropical or temperate, is perhaps, next to that of the palms, the most important feature of the cultivation under glass in the Royal Gardens*'.

Lowe (1895) and Lovis (1967) summarised the early discoveries of reproduction and hybridisation in ferns. They recount how, in 1851, the reproductive strategies of ferns were finally explained by a short-sighted bookseller, F. W. B. Hofmeister. He had spent seven years researching the subject in his own time. By 1863 he was appointed Professor at the University of Heidelberg. Another collection, probably derived from Loddiges, was that of the banker, Robert Barclay. His head gardener, D. Cameron, became curator of the Birmingham Botanic Garden in 1831, and through his influence, it became noted for its richness in exotic ferns. Describing them as '*this long neglected tribe of plants*', Cameron (1828) published a catalogue of the Barclay collection in which the potential of the hardy species, including several native to Britain, was advertised in the leading garden magazine of the day.

The Victorian era was the high point of popularity for ferns. They graced every drawing room, garden and parlour, regardless of class. Glasshouses and conservatories were built for them. Their light, graceful foliage contrasted with the formality of the average home and their elaborate fronds appealed to the Victorian passion for decorating everything. In 1873 B. S. Williams wrote – '*No fete, horticultural exhibition, banquet or public dinner was successful without ferns to grace the occasion, for gay and brilliant colours alone will not satisfy the eyes of the horticultural public . . . Some, indeed, assert that a conservatory properly arranged with ornamental plants and ferns alone, is the most effective*'.

Woolson (1906) romanticised about ferns: '*We have a group of cinnamon ferns before us, the wool of which fairly drips from their uncanny heads like water after a bath, looking very much as if they needed a maternal tongue to lick them into shape*'. He also seemed very unsure about his attitude towards conservation. '*It is worth while to sacrifice specimens from available genera in order to learn how to uproot others..*', and goes on to describe methods for removing ferns such as *Osmunda claytoniana, Dryopteris goldiana, D. spinulosa, Onoclea sensibilis, Dennstaedtia punctiloba, Adiantum pedatum* and *Pellaea atropurpurea* from the wild. But he adds that there's *'no excuse for exterminating rare finds for greed or gain'*.

Botany was considered an acceptable interest for any gentleman, collecting living ferns especially so. Private and commercial collectors scoured the English countryside for the best species and *'hawked them from door to door in their baskets, along with the fishmonger and muffinman'*. This collecting was so effective that many people believe that the countryside has never recovered from this wanton destruction. This led to the first calls for conservation. *'In former days there were not gangs of fern robbers whose everyday business was to practically destroy in any locality that they could find . . . in the Glastonbury District Osmunda regalis is being rapidly destroyed; . . . Ferns are regularly hawked about the streets of our towns by men and even women who are denuding the country of its ferns.'*

Those with the time and/or money joined in the latest pursuit and the ferns stayed in vogue for over fifty years. No other plant has been quite as popular for so long. The previously popular cacti were dismissed by Smith (1866) as *'scarcely saleable'*, and the, then, recently popular orchids as too difficult and too expensive to cultivate. Ferns could be grown, as a general rule, in a comparatively inexpensive manner. In *Fern Growing*, E. J. Lowe (1895) challenged the assertion that ferns could not be crossed by actually crossing them. He divided the prothallus so as to separate the sexes. The idea was to keep it as a prothallus and bring on the sporophyte stage at the will of the grower. He emphasised that fern hunters were not to be confused with the fern robbers. Hunters only chose a single abnormal variety, removed it to a place of safety, and raised the offspring. When two species had been crossed the progeny were more or less sterile, but never completely. The author believed that he was the first to set the ball rolling in the crossing of ferns. Working with Professor Forbes, he attempted crossing experiments. He remembered the disappointment felt when Mr Thomas Moore, of Chelsea Physic Garden, assured him that it was an impossible task. He persisted and carried out mass sowings of mixed plants working on the assumption that, if they were grown touching, minute animal life may assist impregnation. This was successful but very random. There was still great reluctance from botanists to believe that they could be crossed. He undertook painstaking methods to gain recognition. These were carried out mainly between 1857 and 1867. *'..hybrids can be distinguished from crossed varieties in as much as hybrids of species are unproductive, whereas the varieties of species can readily be reproduced by spores.'* By 1867 we had the first printed record of ferns being hybridised, 23 years after the discovery of the reproductive organs. (From Lovis, 1967.)

The Victorians had some unusual ways of dealing with pest problems. *'Under glasses and in cases the green aphis and common scale must be subjected to tobacco fumes, hand-picking and whale oil soap.'* The destructive *'thousand-legged worm'* (millipede) moved so fast that a chase was useless. One had to watch out for him with a pair of open scissors and cut his head off as he ran by. What a picture that must have made! Tiny white worms and black flies in the soil were more difficult to exterminate. A little ammonia or lime in the water had *'a wholesome effect'* and common phosphate sprinkled on top of the soil before watering made them *'gloriously ill'* (Woolson, 1906).

The popular ferns at the time were: the sword ferns (*Nephrolepis*) and their varieties, *Adiantum cuneatum, Pteris cretica* and var. *albo-lineata, Cyrtomium falcatum, Asplenium belangeri* and *A. obtusilobum*, and for the window garden *Lygodium japonicum*. Very popular were *Pteris quadriaurita* var. *argyrea, Blechnum brasiliense, Davallia stricta* and *Polypodium (Phlebodium) aureum*. The fern ball was popular and was made from the hare's-foot fern, *Davallia bullata*, whose rhizomes branch freely and are pliable when wet. They were deftly bound over wire frames filled with *Sphagnum*. The most amusing of these was in the shape of a monkey (Woolson, 1906).

Nathaniel Bagshaw Ward, an eminent surgeon in the 1830s, became famous for being the inventor of the Wardian Case. His character became as much part of the story as his achievements. The vegetable realm was an overwhelming passion for him, and an ambience of greenery almost as essential to him as breath. Demonstrating his concern for the natural world, he called for the sort of action that is only now, 140 years later, gaining popular opinion. '*It is a well known fact, that many hilly countries have been rendered quite sterile, in consequence of the indiscriminate destruction of their trees. . . . Spare the forests, especially those which contain the sources of our streams, for your own sakes, but more especially for that of your children and grandchildren*' (Ward, 1852).

Having inherited his father's practice he went to dingy inner London. Every morning he would rise before dawn to go botanising in the country. In his backyard, in Finsbury Circus, he had a rockery fed by a trickle of water and planted with several species of ferns and mosses. But these failed because the place was '*surrounded by numerous manufactories and enveloped in their smoke*'. He delved into all the sciences and of his hobbies, botany was one and ferns his favourite. He was about to give up his fern ambitions when he was '*led to reflect a little more deeply upon the subject in consequence of a simple incident which occurred in the summer of 1829*'. He recounted '*I had buried the chrysalis of a sphinx (moth) in some moist mould contained in a wide-mouthed glass bottle, covered with a lid. In watching the bottle from day to day, I observed that the moisture during the heat of the day arose from the mould, became condensed on the internal surface of the glass, and returned whence it came, thus keeping the mould in the same degree of humidity. Almost a week prior to the final changes of the insect, a seedling fern and a grass made their appearance on the surface of the mould. I could not but be struck with the circumstances of one of that very tribe of plants, which I had for years fruitlessly attempted to cultivate, coming up spontaneously in such a situation; and asked myself seriously what were the conditions necessary for its growth. To this the answer was – firstly, an atmosphere free from soot; secondly, light; thirdly, heat; fourthly, moisture, and lastly, air.*'

He had a number of small greenhouses designed to his own specifications, made tight with paint and putty and watered only once every five or six weeks. In these he tested plants from a wide variety of families. By mid-1833 he had successfully grown 30 species of ferns as well as the highly challenging bird's nest orchid (*Neottia nidus-avis*). From this Mr Ward's simple bottle was modified, elaborated upon, and became the functional, ornamental and fashionable Wardian Case.

Subsequently, eager fern growers were able to grow a much wider range of species. The poorer grower could afford a smaller case in which to display his collection. '*It is a bit of the woodside sealed down with the life of the wood in it, and when unsealed for a moment it gives forth an odour that might delude us into the belief that we have been suddenly wafted to some dusky dell where the nodding violet grows*' (Hibberd, 1870). Hibberd, an enthusiastic fern grower, wrote many books on the subject. He experimented with growing ferns in glass cases and had some very definite ideas on the art. Some ideas were good, some were disasters. One of his shades '*which was one day removed to a sunny window for a few hours to make room for some domestic operations. The sun heated the air within the shade, the expanded air had no means to escape, and it burst the shade with a loud explosion into a multitude of fragments.*' He thus advised that the glass dome type of case fits loosely into the pan. But, above all, the cases were used by exploring botanists to send new and live plants back to the fascinated public, who flocked to see the new arrivals at Kew.

The development of the Victorian fern craze has been documented by Allen (1969). The simple glass dome was one of the many fern cases used in Victorian times. Glass had become a relatively cheap commodity, and the many inventors of the day constantly came up with better ideas for using it scientifically. In its simplest form, the fern shade consisted of a modified red terra-cotta flower pot, sometimes roughly ornamented, with a wide shallow rim around the top, and a glass dome with a knob on the top. The latter was fitted into the rim which was then filled with water making the whole thing airtight. These were usually no more than 38 cm (15 inches) high, and considered a very neat table ornament. Another form was a glass dish with a similar rim and a bell-glass. Larger versions were produced, some containing ornate pillars with holes around their sides. The latter were filled with damp compost and planted with small ferns and their allies, topped with a cascading fern in a miniature Grecian urn.

Larger and more complex were the aforementioned Wardian cases. Constructed with wood or metal frames and containing boxes or troughs for the growing compost, they were often raised on legs, becoming independent pieces of furniture. Originally simple and rectangular, they became increasingly more elaborate and ornamented to the point that some became miniature Crystal Palaces.

'Miss Maling's' cases were the most popular, having basic rectangular outlines and could, by gas flame or hot water, be heated if required. A patented case had the optional extra of a boiler to supply basal heat, only needing to be filled once or twice a day in winter! The Victorians must have been pretty dedicated. Inside, the cases boasted marvellous miniature landscapes and vistas, with mountains, dells and streams.

Despite universal statements to the contrary, Ward was not in fact the first inventor of these cases. Several others had similar ideas, but did not recognise or capitalise on the future implications. John Ellis came close in suggesting improvements in ocean transport (1771) and William Withering in his *Botanical Arrangement* (1776), a leading manual for British field botanists, came even closer. In 1782 J. E. Smith mentioned a method he had devised for keeping specimens collected for the herbarium fresh for some days, putting their stalks in a small jar of water which in turn was enclosed inside a larger jar, '*closed to prevent the access of air*'.

Unlike these, the invention of a Scottish botanist, A. A. Maconochie, anticipated Ward with quite incredible exactness. About 1825 he became inspired by the work of the Swiss scientist, de Saussure and the great German explorer, Baron von Humboldt. Respectively they showed that plants improved the atmosphere by robbing it of the gases most harmful to animal life, and discovered seaweeds growing, in almost complete gloom, at an immense depth in the ocean. Maconochie (1840) decided that plants living in a fairly densely shaded habitat might well thrive in a confined atmosphere and '*in a light more or less decomposed by the refraction of two surfaces of window glass*'. He then successfully experimented with several exotic ferns and clubmosses in a large glass vessel, and went on to have a '*miniature greenhouse*' made to his requirements. In this he grew, not only ferns, but also orchids and cacti for many years. But, although his friends were familiar with it, he never bothered to make his discovery public until 1840, when Ward's cases had become known. With praiseworthy magnanimity he conceded to Ward all the credit for this outstanding invention.

Meanwhile, Loddiges (see Allen, 1969) had filled some cases with a range of plants and dispatched them to Australia, an idea soon to be copied by Kew. Previously, plants were boxed in moss, or allowed to grow during the voyage with their roots in soil, often thrown about, left unwatered or exposed to the sun, winds and salt-spray. A high proportion normally perished. On their arrival in Australia, the cased plants were in perfect condition.

In March 1834, J. C. Loudon, the proprietor of the enormously popular *Gardener's Magazine* and virtual arbiter of domestic taste to the new middle classes, was invited to visit Mr Ward's residence. He went and was captivated. The house, he eagerly reported to his readers, was literally filled with fern cases. Not all was praise for the Wardian case. '*This very excellent idea, however, was not carried to perfection, for insufficient ventilation, lack of proper drainage and construction, which render filling difficult, may be counted among the defects of the original model.'*

Ward (1852), in his professional capacity as a doctor, saw further implications from his glass cases: *'. . . the air of London, when freed from adventitious matter is as fitted to support vegetative life as the air of the country'*. From this he formed a link between the health of Londoners and the famous smogs of the time. Together with John Lindley, professor of Botany at London University, and several others, he tried to bring influence to bear in the Government to repeal the unjust glass tax. This had made the cost of conservatories virtually prohibitive thus depriving the city's inhabitants of a source of purified air. There was something about the ferns uncannily in tune with the spirit of the age.

They matched the new mood for sombreness; the fern-craze opened as men's clothes suddenly turned black. One writer remarked, with unconscious irony, *'they love a quiet, dull atmosphere'*. Equally ferns suited the growing delight in intricate decoration, with *'the extraordinary exactitudes of their ramifications'*; they formed *'objects of exquisite elegance'*, an inevitable ingredient in Victorian rococo. Above all, they fitted neatly and readily into the content of a romanticism already suffused with the contemporary moral fervour. The Gothic allusions, so long conspicuous by their absence, now began to appear. The popularity of ferns can be judged by the number (over 1,400) of stove, greenhouse, hardy exotic, *Selaginella* and British ferns offered for sale by Birkenhead's nursery in 1890. The 136 page catalogue also had a section on *'hints on the cultivation of ferns'*.

Gas heating and lighting meant the beginning of the end of ferns as indoor plants. The poisonous fumes and dry atmosphere created by the new heating appliances were too much for the tender ferns and they were gradually replaced by the tougher palms, dracaenas, and the ubiquitous *Aspidistra*. Out of doors, though, in graceful little conservatories, contrived grottoes and miniature rockwork landscapes, ferns were still cultivated and much admired.

The turn of the century and the Edwardian era really brought an end to their popularity. Now everything had to be new and severely simple. Ferns were considered far too evocative of the clutter and ornamentation of the Victorian times to go with the new passion for clean lines and uncomplicated designs. The simple palms and aspidistras were much more in keeping with the new fad for elegance.

In a few years, the Great War of 1914-18 was almost to put an end to growing anything more exciting than the basic plants to feed the nation. Great acres of English public and private gardens, once devoted to growing beautiful and exotic plants for pleasure, were turned over to vegetables to feed the people. Many men died in the trenches of the Great War, among them a generation of horticulturalists, collectors, botanists and gardeners. Much skill and a great many plants were lost forever and few nurseries and private gardens were to indulge in fern growing again on the scale of the Victorian era.

Ferns continued as general house plants; their high profile in botanic gardens was slowly taken over by the more showy and 'challenging' plants like orchids. There were no major advances in fern culture at this time. They were, and often still are, propagated conventionally by bulbils, rhizomes, crown splitting, and spores. Wardian cases were replaced by bottle gardens and modern terrariums which enabled the smaller tropical ferns to be grown in the home.

After the veritable plague of horticultural fern books in the late 19th and early years of the 20th centuries, there have been few of note until recent years. Kaye's (1968) *Hardy Ferns* is still *the* reference book for fern gardeners. Bruty (1969) gives useful details for the culture of many greenhouse ferns. Jones (1987) opens up the world of exotic ferns and, with Rush's (1984) *A Guide to Hardy Ferns*, is a most valuable companion to the British Pteridological Society's spore exchange list (which offered nearly 700 species and varieties in 1991). Dyce (1991) keeps up the momentum of the new wave of fern growers with his *The Cultivation and Propagation of British Ferns*.

The latest method of the production of ferns in a carefully controlled environment is that of micropropagation. This relatively new method of propagation involves the growth of very small plant organs, pieces of tissue, seeds or spores in aseptic (sterile) conditions on a nutrient enriched jelly. These are often termed *in vitro* techniques (meaning literally 'in a glass') because the cultures are contained within glass or clear plastic vessels. Ward would have been enthralled! This is usually very much more rapid than traditional propagation methods and has become a powerful tool for studying basic and applied problems in plant biology.

Micropropagation has its roots back in 1927 but was only given its scientific footing in 1934. It was not until the 1960s that this technique really showed its potential when two workers, Murashige and Skoog (1962), developed the now widely used standard media. George and Sherrington's (1984) handbook gives much useful information on this subject.

Ferns are often grown from spores in culture. The resulting gametophytes can then be used for scientific research with possible agricultural applications (such as finding a cost-effective and reliable way to control bracken, *Pteridium aquilinum*, which has been an agricultural problem for over half a century). From this kind of work the horticultural implications were realised. Nurseries require large numbers of uniform plants for their market and because of inherent dangers in conventional propagation of moss and algae invasion, contamination with other ferns, and general slow and uneven growth, micropropagation solved the problem neatly.

Nephrolepis, the popular house plant sold as the sword or Boston fern, is cloned from its rhizomes producing many thousands of plants in a relatively short period of time. *Matteuccia* (ostrich fern), the source of fiddleheads for American consumption, is also mass-produced *in vitro*.

At the Royal Botanic Gardens, Kew, growth from spores *in vitro* is the best way for multiplying ferns that are endangered or difficult to grow (Ford & Fay, 1990). For conservation requirements this is the most desirable method because spores are easier to sterilise than vegetative material and it helps to preserve the genetic composition of the original population. The latter is vital if reintroduction is the final goal. Having plants in aseptic culture has the added advantage of overcoming plant health restrictions for sending plants overseas. There are, on the other hand, disadvantages in that this method may be costly, and there can be rooting and weaning difficulties.

Ferns have had quite a history; the fern has gone from an object of superstition, to *ex situ* propagation in glass cases, to *in vitro* propagation in test-tubes. What the future holds is an unknown quantity but, in the current state of global environmental upset, let us hope that it is good news.

Acknowledgements: I would like to thank Mr John Lonsdale for his help in editing this article and Josephine Camus for her invaluable suggestions.

References
ABBE, E. 1981. *The fern herbal*. New York: Cornell University Press.
AITON, W. 1789. *Hortus Kewensis*. London: George Nicol.
ALLEN, D.E. 1969. *The Victorian fern craze. A history of pteridomania*. London: Hutchinson.
BIRKENHEAD, W. & J. 1890. *Catalogue of ferns*.
BRITTEN, J. 1881. *European ferns*. London: Cassell, Petter, Galpin & Co.
BRUTY, H.J. 1969. The cultivation of greenhouse ferns at Kew. *J. R. Hort. Soc.* **94**: 11-20.
CAMERON, D. 1828. Catalogue of the hardy and exotic ferns in the garden of Robert Barclay, Esq. with directions for their culture. *Gdnrs' Mag.* **4**: 1-6.
DYCE, J.W. 1991. *The cultivation and propagation of British ferns*. London: British Pteridological Society.
ELLIS, J. 1771. Directions for bringing over seeds and plants from the East Indies and other distant countries. London.
EHRHART, F. 1788. Botanische Bernerkungen. *Beitr. zur Natur. Kiel.* **3**: 58-95.
FORD, M.V. & FAY, M.F. 1990. Growth of ferns from spores in axenic culture. *In* Pollard, J.W.

and Walker, J.M. (eds) *Methods in molecular biology.* **6**. *Plant cell and tissue culture*: 171-180. Clifton, New Jersey: Humana Press.
GEORGE, E.F. & SHERRINGTON, P.D. 1984. *Plant propagation by tissue culture. Handbook and directory of commercial laboratories.* Basingstoke: Exegetics Ltd.
HEATH, F.G. 1898. *The fern world.* London: Imperial Press.
HIBBERD, S. 1870. *The fern garden.* London: Groombridge and Sons.
HILL, J. 1769. *Hortus Kewensis.* London: R. Baldwin.
HOLTTUM, R.E. 1967. John Smith of Kew. *Br. Fern Gaz.* **9**: 330-334.
HOOKER, W.J. 1846-1864. *Species filicum* (5 vols) London: Pamplin.
----- & BAKER, J.G. 1868. *Synopsis filicum.* London: Hardwicke.
JONES, D.L. 1987. *Encyclopaedia of ferns.* London: British Museum (Natural History).
KAYE, R. 1968. *Hardy ferns.* London: Faber.
LINDSAY, J. 1794. Account of the germination and raising of ferns from seed. *Trans. Linn. Soc. Lond.* **2**: 93-100.
LOUDON, J.C. 1834. Growing ferns and other plants in glass cases. *Gdnrs' Mag.* **10**: 162-163.
LOVIS, J.D. 1967. Fern hybridists and fern hybridising 1. The work of Edward Joseph Lowe (1825-1900). *Br. Fern Gaz.* **9**: 301-308.
LOWE, E.J. 1895. *Fern growing.* London: Nimmo.
LUO, G.H. 1989. Medicinal pteridophytes in China. *In* Shing, K.H. and Kramer, K.U. (eds) *Proceedings of the International Symposium on systematic pteridology*: 309-312. Beijing: China Science and Technology Press.
MACONOCHIE, A.A. 1840. On the use of glass cases for rearing plants, similar to those recommended by N.B. Ward, Esq. *Rep. Proc. bot. Soc. Edin* **Session 1838-9**: 96-97.
MAY, L.W. 1978. The economic uses and associated folklore of ferns and fern allies. *Bot. Rev.* **44**: 491-528.
MORISON, R. 1699. *Plantarum historia universalis Oxoniensis.* **3**: 554-596. Oxford: Theatro Sheldoniano.
MURASHIGE, T. & SKOOG, F. 1962. A revised medium for rapid growth and bioassays with tobacco tissue cultures. *Physiologia Pl.* **15**: 473-497.
RUSH, R. 1984. *A guide to hardy ferns.* London: British Pteridological Society.
SMITH, J. 1846. An enumeration of ferns cultivated in the Royal Botanic Gardens at Kew in December 1845, with characters and observations on some of the genera and species. *Curtis' bot. Mag.* **72** (Ser. 3: **2**): Appendix 7-41.
---- 1857. *Cultivated ferns.* London: Pamplin.
---- 1866. *Ferns British and foreign.* London: Robert Hardwicke.
---- 1895. *Hand list of ferns and fern allies cultivated in the Royal Botanic Gardens.* London: HMSO.
SMITH, J.E. 1782. *An essay on collecting and preserving specimens of plants.* Smith MSS **39**. Linnean Society, London.
---- 1820. Directions for raising ferns from seed, as practised by Mr. Henry Shepherd of Liverpool. *Trans. R. Hort. Soc. London.* **3**: 338-341.
WANG, DE-QUN. 1989. A tentative investigation of medicinal ferns in China. *In* Shing, K.H. and Kramer, K.U. (eds) *Proceedings of the International Symposium on systematic pteridology*: 313-317. Beijing: China Science and Technology Press.
WARD, N.B. 1852 (2nd ed.). *The growth of plants in closely glazed cases.* Lincoln: Van Voorst.
WILLIAMS, B.S. 1868 (1st ed.), 1873 (2nd ed.). *Select ferns and lycopods.* London: author's publication.
WITHERING, W. 1776. *Botanical arrangement of all the vegetables naturally growing in Great Britain.* Birmingham: Cadel, Elmsley & Robinson.
WOOLSON, G.A. 1906. *Ferns and how to grow them.* London: Heinemann.

Matthew Ford has worked in the Micropropagation Unit at the Royal Botanic Gardens, Kew, for four years. He became interested in ferns during his undergraduate days at the University College of Wales, Aberystwyth and now specialises in the micropropagation of ferns that are rare or difficult to grow, and more recently, woody plants as well. His interests span varied aspects of folklore, ancient customs and folk music, mountaineering and rambling, and photography.

HEXAGONAL WARDIAN CASE.

The Development of Laboratory Based Studies in Fern Variation

Mary Gibby
Department of Botany, The Natural History Museum,
Cromwell Road, London SW7 5BD

One hundred years ago the life cycle of pteridophytes with free living gametophyte and sporophyte generations had been discovered and the deviations of apogamy and apospory recognised, and many pteridologists were attempting to hybridize ferns.

The propagation of ferns from spores was described by Lindsay in 1794, but it was a further fifty years before Nageli (1844) observed the release of spermatozoids from fern prothalli, and Lesczyc-Suminski (1848) discovered the presence of both antheridia and archegonia on prothalli. The following year Hofmeister gave a detailed account of the function of the gametophytes of *Pilularia*, *Salvinia* and *Selaginella*, and he realised that the prothallus in pteridophytes is the morphological equivalent of the leaf-bearing moss plant, whilst the leafy plant of ferns is equivalent to the moss capsule. Henfrey confirmed these discoveries in the true ferns in 1851 [Sachs (1890) provides this historical information].

E. J. Lowe, the author of *Our Native Ferns* (1865), was stimulated after reading Henfrey's paper to attempt to hybridize ferns by sowing dense mixtures of spores. Most of his contemporaries, including the curator of Chelsea Physic Garden and authority on British ferns, Thomas Moore, assured Lowe that this would be an impossible task! However, he persisted, as he was convinced that this would be a new method of raising varieties; he discovered that when two species are crossed they give rise to an *'unproductive'* (sterile) hybrid. He proved to be very successful at hybridizing ferns, and his results are described in his book *Fern Growing, Fifty Years' Experience in Crossing and Cultivation* (1895), which was published not long after the beginnings of the British Pteridological Society; part of this volume he devoted to his theory of multiple parentage, and he believed that he had evidence that more than two parents can combine in a hybrid.

Apogamy, the development of a sporophyte from a prothallus by simple budding, had been described in *Pteris cretica* by Farlow in 1874, and C. T. Druery discovered apospory, the development of prothalli directly from the frond, in *Athyrium filix-femina* 'Clarissima' Jones in 1884.

This was the state of knowledge of the ferns that was described by a vice-president of the British Pteridological Society, C. T. Druery, when he addressed the annual meetings of the Society in August 1894 on *Selective culture of British ferns*, and in August 1895 on *Fern reproduction*. Much had been achieved, despite the fact that Mendel's studies on heredity were not re-discovered until 1900, and chromosomes were only just being observed.

The Society was kept informed of the latest scientific theories at its annual paper reading meetings. Thus, the president, F. W. Stansfield (1896) introduced the members to *Weismann's theory of heredity, and its relation to British ferns* [based on Weismann's book *The Germ-plasm* (1893)], which suggested (i) that the hereditary material lies in the chromosomes within the nucleus (the germ-plasm) and that environmentally determined modifications cannot be transmitted to the offspring and (ii) that two types of cell division would be found to occur, the normal mitosis or equal division (which was then a known fact) and a reduction division, necessary in the production of eggs and sperm so that the original chromosome number would be regained by fertilization.

Strasburger had given an account of mitosis in 1875, and in 1894 demonstrated the numerical reduction in chromosome number during the life cycle of an organism, but it was not until the early decades of the twentieth century that the actual mechanism of meiosis, the reduction division, was accurately described.

Farmer (1894) found that the gametophytic cells of the liverwort *Pallavicinia* have four chromosomes and the sporophytic cells have eight, but fern chromosomes proved to be much more difficult to study, as they tend to have rather high numbers; the early counts are mostly erroneous as the counts were made from cells cut into sections before staining.

In the first years of the twentieth century, the members of the British Pteridological Society were very engrossed in fern varieties; the early numbers of the British Fern Gazette are packed with articles on varieties, and it was in this period that Druery's magnificent book *British Ferns and their Varieties* was published (1910); this includes a chapter on fern hybridizing, and another disputing Lowe's theory on multiple parentage. Druery discussed the new theory of Mendelian inheritance and its application to ferns in a Gazette article (1911); difficulties presented by ferns in the application of Mendel's laws included the discovery that *'sports'* (mutations) are liable to 'sport' again, and the minute size of the reproductive organs *'render crossing on systematic lines a practical impossibility'*. He suggested that Mendel's laws should be tested by crossing normal forms of species, rather than varieties, as these latter proved too confusing. In 1916, Druery questioned why there are so many mutations amongst the ferns compared with other plants – he felt that the usual explanations put forward of changes in the environment, or local differences in environment etc. could not account for the survival of such an array of mutant forms.

Stansfield (1927) gave a report of some very interesting experiments demonstrating Mendelian inheritance in ferns by Miss Irma Andersson (1927) at the John Innes Institute. She sowed spores of a plant of *Polystichum angulare* (= *setiferum*) 'inaequale variegatum', and the following offspring were raised:

158 *P. angulare* 'inaequale variegatum'
 50 *P. angulare* 'congestum'
 63 *P. angulare* 'grandidens'
 14 *P. angulare* 'grandidens congestum',

giving a ratio of approximately 9:3:3:1. The plants of 'congestum' and 'grandidens' were found either to breed true or to give a few plants of 'grandidens congestum'. She suggested that the original plant was *'cross-bred'* (an F1 hybrid), and that two genes were involved, with the 'variegatum' character and that for a normal outline being dominant, and those for 'congestum' and 'grandidens' recessive. Part of her success is attributable to her application of a sterile technique to fern culture. The spores were sown on nutrient-rich agar jelly and the prothallial cultures were clean, with no contamination. She observed the development of the prothallus, with antheridia appearing early, and archegonia developing after the antheridia shrivelled. She even dissected out individual sporangia, and self-fertilized the resulting prothalli, and realised the potential for using such prothallial cultures in hybridization studies.

Allen (1911) made a significant breakthrough in the study of apogamy, when she reported that the number of chromosomes per cell in the prothallus of *Cyrtomium falcatum* was approximately the same as in cells of the young sporophyte (61-62), and that about 64 paired chromosomes were present in eight spore mother cells. Unfortunately her counts were inaccurate, due probably to technical error, although the principle was later demonstrated to be correct. Later, Manton (1950) found *c.* 123 chromosomes from a plant thought to be identical with that studied by Allen. However, Allen made the mistake of believing that the eight spore mother cells arose by fusion from 16 spore mother cells. The full explanation of apogamy in *Cyrtomium* and particularly in *Dryopteris affinis* was clarified later by Döpp (1939) and Manton (1950); they showed that the last mitotic division preceding the formation of the spore mother cells restitutes – the chromosomes divide but the cell does not split into two. This results in the formation of eight, instead of the usual 16 spore mother cells, but each spore mother cell has twice as many chromosomes as the normal sporophytic number; the eight spore mother cells then undergo a normal meiosis and each produces four spores with the sporophytic chromosome number. The resultant sporangium has 32 large spores instead of 64, and the prothalli that germinate from these large spores produce a new sporophyte directly without fusion of gametes.

During the 1930s there was considerable interest by Society members in wild hybrid ferns, with much speculation on which species were the parents of such hybrids (e.g. Kestner, 1935; Stansfield, 1936). Hybrids of *Asplenium* were among those discussed, but perhaps that which aroused the most interest was *Dryopteris remota*. A plant from near Windermere to which this name was applied had been discovered in 1854. Then similar material was reported from Scotland by Boyd (1894) and Ireland by Praegar (1898) (see Manton, 1938). All these forms were compared with similar material that had been found on the continent, and Dr von Tavel from Switzerland (1938) contributed extensively to the discussion of this group of plants. Stansfield (1935) consulted Professor Bower, who suggested that a decisive test of the parentage of Boyd's plant would be to count the chromosomes of the hybrid, and of the putative parents, but *'the cytological experts we consulted think this would be extremely difficult on account of the large numbers of chromosomes'*. At this time few accurate counts of fern chromosomes had been made; Litardière (1921) had discovered polyploidy in ferns although his chromosome counts were not quite accurate but Okabe (1929) obtained correct counts on diploid and polyploid *Psilotum* and Friebel (1933) for *Anemia*. In 1938 a short but important paper by a new member to the Society, Irene Manton, helped to clarify the problem of *Dryopteris remota*. The Windermere plant she found to be sterile, whilst the plants from Scotland, Ireland, and the continent were all fertile; more significantly, she reported that the sterile plant had the same chromosome number as other British species of *Dryopteris*, but that the fertile forms of *D. remota* all had a lower chromosome number.

In 1950, the Rev. Elliot informed members of the Society that Manton had published a book on *Problems of Cytology and Evolution in the Pteridophyta*. This book proved a landmark in the study of pteridophytes. Instead of sectioning cells to study the chromosomes, she applied a new technique of squashing, so that the whole contents of a cell could be viewed in the same plane; application of the squash technique meant that accurate chromosome counts were described for nearly all of the British ferns, and the widespread occurrence of polyploidy and hybridization in the pteridophytes was demonstrated. Manton initiated a research programme that combined controlled hybridization and cytological studies to elucidate the reticulate evolution of various fern genera. Many of the results from the research projects of Manton's students were published in the Gazette during the 1960s, for example Shivas (1962) on *Polypodium*, Lovis (1963, 1964) and Sleep (1967) on *Asplenium*, and especially in the 1969 volume that was dedicated to Professor Manton on her retirement from the Botany Department at Leeds – Lovis and Vida on *Asplenium*, Shivas also on *Asplenium*, Braithwaite on *Hymenophyllum* and Walker on *Dryopteris*. During a study of *Asplenium aethiopicum*, Braithwaite (1964) discovered a new type of apogamy in ferns, where the restitution occurs during meiosis. Similar work on fern evolution was continued in the next two decades, by these same authors, and by a second generation of students.

At the same time, rapid developments were being made in the use of chemistry in fern classification (Cooper-Driver & Haufler, 1983). Widén demonstrated the application of the study of aromatic compounds, the phloroglucinols, for further understanding of systematic relationships in *Dryopteris*; for example, eastern and western American forms of *D. expansa* could be distinguished (Widén & Britton, 1969), and the parentage of *D. remota* (= *D. expansa* x *D. affinis*), proposed on morphological and cytological evidence, was confirmed (Widén et al., 1971). *Asplenium cuneifolium* can be separated from *A. adiantum-nigrum* and *A. onopteris* on the basis of its polyphenolics (*C*- glycosylxanthones) (Richardson & Lorenz-Liburnau, 1982). Study of protein chemistry has proved valuable for investigations both between and within species. Protein molecules carry a small charge and so will tend to migrate in an electric field. Proteins of different charges and sizes can therefore be easily separated. Different species tend to have different forms of the same protein, and such differences can be used to compare variation in enzyme patterns between, for example, a polyploid and its proposed diploid progenitors (Haufler, 1985a, on North American *Cystopteris*), or to study variability between populations or even individuals (Haufler, 1985b). Now chemical analysis has progressed to the level of studying the actual material of inheritance, the DNA. A special class of enzymes, restriction enzymes, cut DNA at specific points into small fragments. Like proteins, DNA fragments are charged and so can be separated by size in an electric field. The pattern of different sized fragments is often characteristic of different taxa. By comparing restriction fragment length polymorphisms in chloroplast DNA, Stein and Barrington (1990) have shown that hybrid individuals in a population of *Polystichum* x *potteri* (= *P. acrostichoides* x *P. braunii*) had chloroplast DNA like one or the other parent. This suggests that the hybrids in this population have arisen more than once, as chloroplast DNA is believed to be inherited only via the 'female' line.

A hundred years ago pteridologists had only the acuity of their own sight to describe and define the variation before them. The intervening period has seen variation described at progressively finer and finer levels, first of all the chromosomes, then secondary chemical compounds, then proteins and now the sequence of the code of DNA itself. For the future, DNA studies will prove invaluable, not only for studies between related species, but also for investigations at the generic level, and the place of ferns in the evolution of the Plant Kingdom will be more firmly established. But despite the application to ferns of the techniques that modern biology can provide, pteridology, like astronomy, remains a field in which the amateur can still make valuable contributions.

References

ALLEN, R.F. 1911. Studies in spermatogenesis and apogamy in ferns. *Trans. Wis. Acad. Sci. Arts Lett.* **17**: 1-56.

ANDERSSON, I. 1927. Note on some characters in ferns subject to Mendelian inheritance. *Hereditas* **9**: 157-168.

BRAITHWAITE, A.F. 1964. A new type of apogamy in ferns. *New Phytol.* **63**: 293-305.

---- 1969. The cytology of some Hymenophyllaceae from the Solomon Islands. *Brit. Fern Gaz.* **10**: 81-91.

COOPER-DRIVER, G.A. & HAUFLER, C. 1983. The changing role of chemistry in fern classification. *Brit. Fern Gaz.* **12**: 283-294.

DÖPP, W. 1939. Cytologische u. genetische Untersuchungen innerhalb der Gattung *Dryopteris*. *Planta* **29**: 481-533.

DRUERY, C.T. 1884 Observations on a singular mode of development in the lady fern (*Athyrium filix-foemina*). *J. Linn. Soc. Bot.* **21**: 354-357.

---- 1894. Selective culture of British ferns. *British Pteridological Society, abstracts of reports* **1894-5**: 5-9.

---- 1895. Fern reproduction. *British Pteridological Society, abstracts of reports* **1894-5**: 20-25.

---- 1910. *British ferns and their varieties*. London: George Routledge and Sons.

---- 1911. Ferns and Mendelism. *Brit. Fern Gaz.* **1**: 245-246.

---- 1916. The cause of variation. *Brit. Fern Gaz.* **3**: 63-67.

ELLIOT, E.A. 1950. Editorial. *Brit. Fern Gaz.* **7**: 265.

FARLOW, W.G. 1874. An asexual growth from the prothallus of *Pteris cretica*. *J. Microscop. Sci.* (new ser.) **14**: 266-272.

FARMER, J.B. 1894. Studies in the Hepaticae: On *Pallavicinia decipiens*. *Ann. Bot.* **9**: 35-52.

FRIEBEL, H. 1933. Untersuchungen zur Cytologie der Farne. *Beitr. Biol. Pfl.* **21**: 167-210.

HAUFLER, C. 1985a. Pteridophyte evolutionary biology: the electrophoretic approach. *Proc. Roy. Soc. Edinburgh* **86B**: 315-323.

---- 1985b. Enzyme variability and modes of evolution in *Bommeria* (Pteridaceae). *Syst. Bot.* **10**: 92-104.

KESTNER, P. 1935. More on the Brissago hybrid and on hybrid ferns in general. *Brit. Fern Gaz.* **7**: 19-24.

LINDSAY, J. 1794. Account of the germination and raising of ferns from the seed. *Trans. Linn. Soc. Lond.* **2**: 93-100.

LITARDIÈRE R. de. 1921. Recherches sur l'élément chromosomique dans la caryocinèse somatique dans Filicinées. *Cellule* **31**: 255-473.

LOVIS, J.D. 1963. Meiosis in *Asplenium x murbeckii*. *Brit. Fern Gaz.* **9**: 110-113.

---- 1964. The taxonomy of *Asplenium trichomanes* in Europe. *Brit. Fern Gaz.* **9**: 147-160.

---- & VIDA, G. 1969. The resynthesis and cytogenetic investigation of x *Asplenophyllitis microdon* and x *A. jacksonii*. *Brit. Fern Gaz.* **10**: 53-67.

---- 1865. *Our native ferns*. London: George Bell.

---- 1895. *Fern growing/ fifty years' experience in crossing and cultivation/ with a list of the most important varieties and a history of the discovery of multiple parentage, etc.* London: Nimmo.

MANTON, I. 1938. Hybrid *Dryopteris* (*Lastrea*) in Britain. *Brit. Fern Gaz.* **7**: 165-167.

---- 1950. *Problems of cytology and evolution in the Pteridophyta*. Cambridge: Cambridge University Press.

OKABE, S. 1929. Ueber eine tetraploide Gartenrasse von *Psilotum nudum* Palisot de Beauvois (= *P. triquetrum* Swartz). *Sci. Rep. Tohoku Univ.* (ser. 4.) **4**: 373.

RICHARDSON, P.M. & LORENZ-LIBURNAU, E. 1982. C-glycosylxanthones in the *Asplenium adiantum-nigrum* complex. *Amer. Fern J.* **72**: 103-106.

SACHS F.G.J. von. 1890. *History of botany, 1530-1860*. (translated by H.E.F. Garnsey; revised by I.B. Balfour). Oxford: Clarendon Press.

SHIVAS, M.G. 1962. The *Polypodium vulgare* complex. *Brit. Fern Gaz.* **9**: 65-70.

---- 1969. A cytotaxonomic study of the *Asplenium adiantum-nigrum* complex. *Brit. Fern Gaz.* **10**: 68-79.

SLEEP, A. 1967. A contribution to the cytotaxonomy of *Asplenium majoricum*. *Brit. Fern Gaz.* **9**: 321-329.

STANSFIELD, F.W. 1896. Weissmann's theory of heredity and its relation to fern life. *British Pteridological Society, abstract of report* **1896**: 10-21.

---- 1927. Mendelism in ferns at the John Innes Institute. *Brit. Fern Gaz.* **5**: 165-170.

---- 1936. More about hybrid ferns: a review of the situation. *Brit. Fern Gaz.* **7**: 86-92.

STEIN, D.B. & BARRINGTON, D.S. 1990. Recurring hybrid formation in a population of *Polystichum* x *potteri*: evidence from chloroplast DNA comparisons. *Ann. Missouri Bot. Gard.* **77**: 334-339.

STRASBURGER, E. 1875. *Ueber Zellbildung und Zelltheiling*. Jena: Hermann Dabis.

---- 1894. The periodic reduction of chromosomes in living organisms. *Ann. Bot.* **8**: 281-316.

TAVEL F. von. 1938. Hybrid *Dryopteris*. *Brit. Fern Gaz.* **7**: 134-135.

WALKER, S. 1969. Identification of a diploid ancestral genome in the *Dryopteris spinulosa* complex. *Brit. Fern Gaz.* **10**: 97-99.

WEISMANN, F.L.A. 1893. *The germplasm: a theory of heredity*. (translated by W.N. Parker & H. Ronnfeldt etc.) London: Walter Scott.

WIDÉN, C.-J. & BRITTON, D.M. 1969. A chromatographic and cytological study of *Dryopteris dilatata* in eastern North America. *Can. J. Bot.* **47**: 1337-1344.

---- VIDA, G., EUW, J. von & REICHSTEIN, T. 1971. Die phloroglucide von *Dryopteris villarii* (Bell.) Woynar und anderer Farne der gattung *Dryopteris* sowie die mögliche Abstammung von *D. filix-mas* (L.) Schott. *Helv. Chim. Acta* **54**: 2824-2850.

Mary Gibby's early interest in botany was further inspired by her undergraduate days with Prof. Manton at Leeds University and she took up the study of ferns and chromosomes. Her doctoral research on the genetics of Dryopteris was carried out with Dr Stanley Walker who had been Prof. Manton's first research student at Leeds. She pursues her career at the Natural History Museum, London, with studies of the cytology of ferns, and more recently, pelargoniums.

ASPLENIUM NIGRUM GRANDICEPS.

ASPLENIUM TRICHOMANES

ASPLENIUM TRICHOMANES CRISTATUM.

The Story of the Reginald Kaye Fern Collection

R. Kaye

Waithman Nurseries, Silverdale, Carnforth, Lancs. LA5 0TY

Really the story of my interest in ferns dates back to my later schooldays though it was not until Waithman Nurseries, which I started in 1930, had been established some three or four years that I began to build up a fern collection in Silverdale.

As a schoolboy I spent many holidays with my grandfather who lived in Harrogate, and I remember he had a small tufa rock garden covered chiefly with *Polystichum setiferum* var. *acutilobum*. This fern always fascinated me by its habit of covering its rachises with masses of bulbils. In a lane behind my grandfather's property there was a very good alpine plant nursery where in a large rock garden Alva Hall grew a large collection of alpine plants to perfection. In the shadier parts thereof he had several specimens of varieties of *Athyrium filix-femina* with its wide range of crested, cruciate, and other forms, interspersed with cyclamens, trilliums and other treasures. These ferns were a revelation to me and whetted my appetite to grow such fascinating plants.

In those days I lived in Huddersfield with my mother who was a keen gardener, though she could never get the names right. We frequently went over to Brighouse to Kershaw's old nursery – now no more – and on one occasion I obtained there a fern labelled *Osmunda gracilis*, a form of *O. regalis*, with tall slender fronds, copper tinted in spring. Some seventy years later, I still grow the original plant at Silverdale. Unfortunately this variety seems to be sterile, although it produces masses of spores. These are pale cream in colour through chlorophyll being absent, and repeated sowings have produced nothing; it can only be increased by division. About this time, I purchased a copy of Druery's *British Ferns and their Varieties* and this work became my bible for many years. However, I never came across any more fern growers until, when I was about twenty, some school friends formed a party addicted to walking and exploring the Yorkshire Dales, and ultimately, Lakeland. Here in Borrowdale I discovered William Askew's Fern Nursery, which was very extensive, being two acres. Almost all of this was devoted to a very comprehensive collection, mostly in fine condition. A considerable part of the nursery was overrun by *Tropaeolum speciosum* which completely smothered some of the ferns. Mr Askew used to sell a bucketful of *Tropaeolum* tubers for one shilling. He would only sell the commoner fern varieties. Amongst other things there was a splendid collection of *Polypodium* variations, but getting any of the less common ones from him was like drawing teeth.

I was training as an analytical chemist in those days, but after two years spent as assistant to the Borough Analyst for no pay, and fruitless searching for jobs, I decided to grow alpines and lay out gardens from my mother's home, and got quite busy. One day at the Chelsea Show, I was introduced to Walter Ingwersen who invited me to join him at Gravetye. So I packed up all my alpines and ferns – including the *Osmunda*, and got a house in East Grinstead – I also got married at the same time – and enjoyed three years in Sussex. But the pull of our northern hills and clear mountain streams haunted me until I decided to return north and start my own place. Three months of looking for a nice spot with land available finally led me to Silverdale where I decided to settle – I think the local limestone went to my head. (And the *Osmunda* came too.)

I joined the British Pteridological Society sometime in the later 1920s, when I met many of the older fern enthusiasts including some of the founder members of the Society. Amongst other memories I recall visiting the 'old' Dr Stansfield in Reading. Though he had quite a small garden, it contained many treasures. I remember seeing *Polypodium vulgare* 'Grandiceps Parker' which is illustrated in Druery's book. Dr Stansfield kindly presented me with an inch or so of the rhizome which I carefully tended, but alas, it never made a growing point. I believe that this fern no longer exists.

Soon after I came to Silverdale, I met J. Barnes, the headmaster of Earnseat School in the grounds of Ashmeadow, an extensive woodland garden on a north-facing slope just behind the sea wall. In those days there was an extensive collection of ferns. Mr Barnes was the son of J. M. Barnes, the famous collector of many fine varieties of *Lastrea propinqua*, now *Dryopteris oreades*, but even then nearly all of them had disappeared from cultivation. I acquired my first 'Bevis' as a gift from Mr Barnes.

Mr Barnes, then quite elderly, was kept very busy running the school and had limited time for looking after the ferns. I remember he had a large bed of *Polypodium vulgare* 'Omnilacerum Moore'. On a later visit the bed was empty. The polypodies die down in spring and the new fronds seldom appear before midsummer, and the gardener must have dug the bed over to keep the place tidy. By the way, I am using the names extant at that time which was before the species *australe* [= *cambricum* Ed.], *interjectum* and *vulgare sensu stricto* came into use. I remember there used to be a long border, some forty feet by five, which was full of fine varieties of hart's tongue – then *Scolopendrium vulgare*. I have not been to see Ashmeadow for some years but I understand that even this bed has gone.

I became very friendly with Fred Jackson of Borrowdale, who had a good collection in his garden at Stonethwaite. One of Fred's specialities was a collection of *Asplenium* species and hybrids which he used to grow in pans sunk in sphagnum moss. Fred was a real enthusiast and I visited his garden many times. Fred came to Silverdale where he helped me to make some raised beds with stone walls. One bed about forty feet long with rounded ends we christened 'Fred's Barrow'. The theory was that on their deaths Fred should be buried at one end, James Davidson at the other end, and yours truly down the middle, each with a bottle of his favourite brand of alcohol between his knees. Alas both Fred and James have died and have not taken advantage of the plan.

When Fred was getting older he went to live with his daughter on the Isle of Cumbrae, and he had most of his pet ferns sent up to Scotland where, I am told, they still thrive. He left the *Asplenium* collection behind, forty pans of them. I fear that they were sadly neglected and once when I called I found that they had not been watered for weeks, so I got Fred's permission to take them all to Silverdale and try to bring them round. Out of the whole lot of forty pans only four recovered. These were divided up and potted in thumb pots and when they were growing freely I sent all except one or two up to Fred in Scotland. *Asplenium* x *alternifolium*, *A.* x *murbeckii* and *A. trichomanes* 'cristatum' were the survivors and have made nice plants in my raised fern bed.

A relation – whose name I cannot recall – brought to my notice another collection, that of Percy Greenfield. This was lined out in the garden of a delightful old cottage at Bletchley, Buckinghamshire. There were about fifty old clumps of *Polystichum*, all without labels. The place seemed deserted but I finally made contact with the owner who was not at all interested in the ferns and I bought the lot for a very reasonable amount. I lifted them and packed them off to Silverdale in a series of tea chests. Amongst them were two large clumps of 'Bevis' which when divided made about sixty nice plants and a fine clump of *Polystichum setiferum* 'Multilobum' – a very handsome 'Divisilobum'. I lined out the collection and got W. B. Cranfield, who was attending a week's fern hunting jaunt based on Kendal, to come round to assist me in putting names to them. Mr Cranfield invited my wife and me to spend a couple of days with him at Enfield where there was the most comprehensive set of British fern varieties ever amassed. Mr Cranfield was getting on and offered me the entire collection for £500, but at that time I had a most unsympathetic bank manager and I felt unable to accept the offer. Mr Cranfield ultimately left the collection to the Royal Horticultural Society's garden at Wisley, and that was the end of that collection.

The Reginald Kaye Fern Collection

My old friend Robert Whiteside was one of the earliest members of the BPS, and had built up a fine collection in a large garden in Lancaster, but after moving to a smaller house in Bare, Morecambe, was feeling the effects of growing old at 94. He still maintained his collection but it was becoming a burden and he asked me if I would like to buy them at a very moderate figure, to which I agreed. I dug up three lorry loads and transferred them to Silverdale. Unfortunately the fence separating his garden from neighbouring fields had broken down and some horses had been enjoying themselves rampaging up and down. All the labels had been kicked out and broken, and many were illegible, so getting all the names right was a problem even with Robert's assistance as many of the fronds were damaged.

One particular plant which Robert regarded as his best wild find, *Dryopteris dilatata* 'Crispa Whiteside', was a mere stump. I planted it out with a large glass jamjar over it. Nothing happened the first year, the following year one small frond appeared, and a year later I moved it to a good position where after four years it made normal growth. A few years later it suddenly produced masses of fertile fronds and I was able to raise about four hundred plants which were all true to type when nursed to full size. I remember Clive Jermy called in about that time and he was most impressed with it. This year it has produced a lot of spores again after being barren for several years.

Then there was a Mr Penny who also lived in Lancaster and also moved to Morecambe, taking his ferns with him however. He came to see me one day and offered me his collection at a nominal sum. He said it was hopeless trying to keep his treasures in good shape as his garden had been adopted by all the neighbourhood's feline population who used his garden as a meeting place for their midnight orgies, so I moved them to Silverdale. They mostly were very nice athyriums, one of which, a very good 'Plumosum' he had named 'Penny's' variety.

Ray Coughlin, who has a fine fern collection near Bromsgrove had *Athyrium filix-femina* 'Plumosum Penny' from me some years ago. He raised a beautiful variety 'Plumosum Kalothrix' whose fronds really suggest silken tresses, and sporelings from this come true to variety. There was an old variety under this name but it disappeared from cultivation many years ago. Mr Coughlin's plant is not identical with the old 'Kalothrix' but well deserves the name. In pre-war days Charlie Grubb had a nursery at Bolton-le-Sands which originally was devoted to tomatoes, chrysanthemums and the like, but when Charlie became interested in ferns he really set about raising stocks by the thousand by sowing spores. He built up a fine stock almost explosively, and many excellent plants were on display. I spent many a visit there and had many good plants from him, but when Charlie died the nursery became a building site and is no more.

In addition to the above I further increased my collection when John Stormonth died. John was one of the most knowledgeable plantsmen I ever met. His nursery was started in the later 1800s on five acres at Kirkbride near Carlisle, and might well have been mine for when I was looking round for a place to start my own, John Stormonth was thinking of retiring. After some correspondence I spent a day at the nursery and agreed on a price for the nursery and house, but then John decided not to sell and carried on. He said that he doubted if I would succeed unless I was prepared to work day and night and he would hate to see the place go down. Actually he could not bear the idea of leaving his life's work. However, John died and to begin with I was offered the stock which I bought and started removing plants down to Silverdale. But after I had removed a small section, Miss Stormonth, John's sister, decided to sell out altogether and I took over the nursery and put my foreman from Silverdale in charge. I made weekly visits to check progress. There was a well established seed business and a large seed warehouse which almost ran itself as I took over John's staff as well. The nursery was a treasure house of rare plants. For instance, I had already taken to Silverdale the complete stock of fifty named varieties of *Helleborus orientalis*, about a score of each kind. I also had taken John's extensive fern collection to Silverdale. This was in

1937. I also acquired eight hives of bees with the Stormonth nursery which were kept mainly to pollinate the three acre orchard which was underplanted with shade-loving plants, so I had to learn bee-keeping and got very interested. By this time I had amassed probably the best collection of hardy plants in the country. I had a staff of ten in the nursery, and two in the office at Silverdale. Amongst other work I did a good deal of laying out of rock gardens up and down the country in the autumn and winter. One job was at Broughty Ferry, near Dundee, which only allowed brief weekend returns to home. While I was away the man in charge of the nursery sold the entire stock of hellebores to a visitor at two shillings and sixpence a plant cash, and no record of his name was taken. My foreman seemed to think he had been smart and was quite upset when I lost my temper with him. I have never been able to replace those hellebores. By this time I had amassed a fine collection of ferns: about five hundred different varieties of British ferns plus quite a few foreign ones.

Then the war came along and while my staff at Kirkbride remained intact, by the end of the war I was left single-handed at Silverdale for six months after a procession of land girls, and during which time our main preoccupation was growing vegetables. Apart from the local sales, our branch, John Stormonth & Son, had two permanent stalls at Carlisle market through which I managed to clear most of the vegetables we grew. I can remember getting up at 4 a.m. to pick half a hundredweight of fine green peas and getting them to Kirkbride in time for the Friday market.

During this period the fern collection had to look after itself, which it did remarkably well apart from getting weedy, and after getting the last of the vegetables away we had a good deal of vacant land in good heart which we soon filled up with herbaceous plants, shrubs and ferns. But it took some years before we even approached our pre-war collection, as many species had been lost altogether. We have never recovered our former stock list.

I was away from home a good deal making rock gardens but managed to build up the fern collection from sowings of spores at weekends. I would have liked to transfer my nursery to Kirkbride but the idea upset my wife and I decided to sell the Kirkbride nursery and concentrate on rebuilding the Silverdale business. My foreman at Kirkbride expressed a wish to take over rather than return to Silverdale and I agreed on satisfactory financial arrangements (very sympathetic ones too). However, he went bankrupt after three years and the nursery is now built over.

In 1968 I published my book *Hardy Ferns* which was well received and created a much greater demand for ferns than in earlier years, but we managed to propagate good stocks. By this time I had worked up quite a nice collection of New Zealand ferns. A correspondent in the South Island frequently sent me small parcels of nice things which we propagated until we had filled a forty-foot heated greenhouse, as most of the species were not frost-hardy. About this time I went into hospital for a hip replacement which is still going strong, but while I was away vandals amused themselves by throwing top-stones off our boundary wall through the roof of the greenhouse which ran parallel to the road. A spell of hard frosts occurred at the same time, and when I got back from hospital most of the New Zealand ferns were defunct.

Over the years many of the ferns had lost their labels, and while I could put a name to most of them, others were uncertain. Jimmy Dyce and Bert Bruty (from the Royal Botanic Gardens, Kew) came to Silverdale for a week to try and sort out the polystichums of which I had thirty differing forms of *P. setiferum* 'Divisilobum', some of them original clones from the Whiteside collection. So fronds were taken for the herbarium all numbered, and Bert took some packets of Hartley labels to write down their recommended names with a new pencil said to be quite weatherproof. Unfortunately, no one thought of numbering the ferns in the nursery to correspond with their findings and I left the labels in the ground in the fern garden. Alas, within a month the letters were reduced to pale grey blobs, quite illegible. Moral - use an engraving tool next time. About that time some of the old plants 'went off' through urgent need of replanting – too many things to do as usual. So I have lost some of the plants over the years. Still, fresh ones keep turning up and some very nice things have appeared during the last year or two.

Two years ago I made the property and business over to my son and his wife, and I think he will keep the collection going. At 88 I try to keep all the ferns in good condition and labelled with engraved Hartley labels – which keep turning up when digging.

In recent times I have had some very nice additions to the collection from Martin Rickard, with whom I have lengthy telephone conversations. I am writing a revised book on hardy ferns at present and I get confused with the change over of many varieties to cultivar status. Some years back I had a very nice sport appear which I named provisionally *Athyrium filix-femina* 'Crispum Grandiceps' and I sent a frond to Prof. R.E. Holttum for his opinion. He advised me that I should give it a cultivar name so I wrote suggesting 'Baby Bighead' as appropriate, but I never received a reply.

One of the headaches is the present adoption of specific status for the plumose sport of *Polypodium australe*, at present named *Polypodium cambricum*. But there are several plumose 'Cambricums' such as 'Barrowii', 'Hadwinii', 'Oakleyi', 'Prestonii', and 'Wilharris' all in cultivation. I understand that there is some extensive research on *Polypodium* in the United States and one hopes that some of the anomalies may be ironed out. In my first book on hardy ferns, all the varieties were described as belonging to *P. vulgare*. At that time the division of *P. vulgare* into three subspecies (now species) was only just coming to notice. Now it seems that very few varieties are attributed to *P. vulgare sensu stricto*.

Not long ago I was invited to designate my collection as 'National' for the purpose of conservation. But this involved preparing maps of all the many beds to show the position of each variety and the beds are being constantly reorganised with varieties moved about in the attempt to get rid of weeds, so it would not be very convenient to keep such maps up to date. Paths would have to be kept spotless, and free access provided to all and sundry. After due consideration I decided that I must decline. I shall endeavour to maintain the collection as well as I can, and when I am no longer able to look after my ferns, my son will take over. I am always pleased to take interested visitors round the nursery.

FERNERY IN FORM OF A MOUND.

Conservation – The Fern Story

Martin H. Rickard

The Old Rectory, Leinthall Starkes, Ludlow, Shropshire SY8 2HP

During the nineteenth century the pursuit of the picturesque was reaching a climax, garden structures, such as grottoes and follies, were highly fashionable; ferns were ideal for culture in many of these. In about 1840 Nathaniel Ward developed his Wardian Case idea – allowing the growth of ferns in the most polluted town environments. Fern popularity boomed, dozens of books were published and fern motifs appeared on almost every conceivable type of household item, many of them highly collectable today.

The first known attempt at fern conservation was not, however, until around 1880, when E. J. Lowe petitioned the Board of Works to get a fernery built at Kew to house a National Collection of Ferns (Lowe, 1895). This would enlarge the hardy fern collection established at Kew in 1874. Lowe, Dr E. F. Fox and Col. A. M. Jones agreed to supply plants and in 1887 the collection was greatly enlarged by the addition of 4,261 plants willed to Kew from W. C. Carbonell's collection at Usk in Monmouthshire (Anon., 1895). Thus, a National Collection of around 5,000 ferns, but not all different, was set up at Kew over 100 years ago.

We read little about the Kew collection until a passing reference (Druery, 1915) makes it clear that, although the collection was still enormous, it was not of the highest quality, and large numbers of unnamed seedling ferns had become established. In 1925 three of the leading fern growers of the day, W. B. Cranfield, T. E. Henwood and Dr F. W. Stansfield, visited Kew on the invitation of the Director, Dr A. W. Hill, for the purpose of overhauling and reorganising the collection (Stansfield, 1925). At this date there was still a considerable number of plants in the collection, but many were duplicates or indefinable forms of *Polystichum setiferum* 'Acutilobum' or 'Divisilobum'. Although most were probably seedlings there were still some good old standard varieties, e.g. *Polystichum setiferum* 'Tripinnatum Padley', 'Latifolium Moly', 'Setoso-cuneatum Phillips', 'Revolvens Wills', 'Grandiceps Talbot', 'Grandiceps Abbot', 'Venustum Padley' and 'Multilobum Ovale'. None of these are known by name today. There was also a fine patch of *Dryopteris filix-mas* 'Bollandiae'; today this variety has become rather rare but happily it still prospers at Kew. Other genera of ferns were even then poorly represented.

In 1925 it became clear that Kew was not climatically the ideal site for a collection of ferns – pollution had increased since 1880 and the rainfall was, of course, too low. Accordingly, it was proposed to keep the more choice varieties under glass, and the British Pteridological Society at this time undertook to supply plants of a range of standard varieties to try and build up the collection again to the desired standard of a National Collection.

Today there are still hardy ferns at Kew. As well as *Dryopteris filix-mas* 'Bollandiae', mentioned above, there are some very fine *Polystichum setiferum* varieties especially 'Divisilobum'. However, this collection falls far short of being a show-piece for the wonderful range of forms shown by our native British ferns, and it would be inappropriate now to call it a National Collection.

Also during the 1880s, Col. A. M. Jones donated the greater part of his superb collecton to the Bristol Zoological Gardens (Lowe, 1895). Here, they thrived for many years, certainly in fine form until 1909 at least (Cranfield, 1910). I have never visited these gardens but I gather that there are few, if any, first rate varieties there now.

A little later, in the 1880s or 1890s, the fernery at the Royal Botanic Gardens, Glasnevin, Dublin was expanded; this too was supported by E. J. Lowe (Lowe, 1895). I do not think it was ever a very comprehensive collection but over the years it was no doubt supported by notable Irish growers such as W. H. Phillips, Canon H. Kingsmill-Moore, P. B. O'Kelly and R. L. Praeger. Today I believe few fern varieties survive although native species are well represented, including the protected Killarney Fern (Charles Nelson, pers. comm., 1988).

Shortly after the establishment of these Victorian collections, in 1891, the British Pteridological Society (BPS) was formed. This was the second attempt at forming a national society devoted to the culture of ferns. The first lasted only two or three years during the 1870s. Although short-lived, this older society did succeed in promoting the establishment of these early fern collections as well as stimulating the publication of the *'Jones' Nature Prints'* (Jones, 1876-80) – probably the most valuable publication on British fern varieties ever produced (Dyce, 1989).

Through the closing years of the nineteenth century and up to the First World War the BPS provided the forum for growers to meet and exchange choice plants. The membership remained small (about 100) but many members held good representative collections, most of which, I suspect, would have sent any one of today's growers into raptures if they could be seen today! By the time of the First World War many members were too old to become involved in the hostilities and were therefore able to continue tending their collections. However, new pteridophiles were few and far between and by the 1920s the membership of the BPS was dwindling as the old pioneers died one by one and their collections were dispersed or sold to the highest bidders. One member often seemed to get the best collections – W. B. Cranfield, the President of the BPS from 1920 to 1948. He was a rich man with a four acre garden who could afford to buy ferns without financial restriction and be sure of the space to grow them. Eventually Cranfield built up a collection better than any other that had gone before. Unfortunately, he does not seem to have been very generous with his spare plants, having the sole representatives of many varieties, but on his death in 1948 he willed his collection to Wisley. Just post-war, Wisley was, no doubt, unprepared for such a vast collection of ferns and only one van was sent to pick up just one load. Therefore, only a fraction was collected and the new owners of the estate soon burnt off the remainder and planted the ground with vegetables.

At a stroke the vast majority of the last great fern collection was destroyed. Those plants which made it to Wisley did not seem to fare much better; some no doubt survived, but it was not until quite recently that any sort of fern collection could be seen there. This was by courtesy of Graham Stuart Thomas who presented a small, yet very choice, range of ferns. when I last saw it in 1987 I was amazed to see several plants of the very rare *Polystichum setiferum* 'Plumosum Grande Moly' but sad to see the dense weed cover which was developing; in places cleavers (*Galium aparine*) was head high. I believe this collection has now been moved to another part of the garden where it should prosper.

Fortunately, Cranfield did not always get all the ferns in circulation. Some collections were passed down by gift, or perhaps by other routes, to younger members of the BPS and today it is these few enthusiasts we have to thank for most of our best ferns in cultivation. In the south of England F. W. Stansfield left many of his ferns to his son-in-law, Percy Greenfield, who passed many on to their present owner – J. W. Dyce. Jimmy Dyce has a particularly prominent position in the history of fern cultivation this century. Almost single-handed he resurrected the BPS after the Second World War and served in almost every one of its offices at one time or another. He has in fact been an officer of the Society since 1939 – today he is the Society's first ever President Emeritus and, as ever, is still generously distributing ferns to keen younger members of the Society. Another key member who joined the BPS in the 1930s is Jean Taylor, now Jean Healey, who started accumulating material for her nursery at Bracknell. Recently, most of her ferns were dispersed but this collection was an important modern source of some choice varieties.

Northern collections were also changing hands, some being preserved by the Brookfield family and many more by Reginald Kaye at his Silverdale nursery. Reginald Kaye, a member of the BPS since 1929, bought Robert Whiteside's entire collection and odd plants from the collections of J. Barnes, Cranfield, Stormonth, Grubb and others, as well as a large collection from a garden in Bletchley.

There were, of course, other important collections, e.g. the Dyer collection moved by A. R. Busby to the Oxford Botanic Gardens during the 1970s, but one very important collection is worthy of special mention here – the collection of the Bolton family. This was initiated during the heyday of the fern craze in North Lancashire, and much of it still survives today within the family. One large collection now held at Austwick in Yorkshire is still a source for reintroducing much choice material, while the nucleus of another collection is still cared for at Birdbrook in Suffolk where the family run their well-known Bolton's Sweet Pea business. The generosity of the owners of the Birdbrook ferns is legendary. Members of the BPS were once issued with garden forks during a visit and offered anything they wanted as long as it was not the only specimen of a particular variety. Amazing generosity like this has resulted in the perpetuation of many of our finest fern varieties.

Today, there are, therefore, still some fine fern varieties in cultivation, although many have been lost. One hundred years ago Lowe listed and described 1,859 varieties of British ferns in *British Ferns and Where Found* (Lowe, 1890). Although this work was a slim volume which took only six weeks to write it remains the most comprehensive listing of fern varieties ever published. Subsequently, of course, many more fine varieties have been found or raised, making in all a total of, perhaps, 3,000 varieties of British ferns which have been described to date.

Obviously, I cannot report that all of these varieties are safe and well, thriving in gardens scattered throughout these islands. The most recent account of hardy fern varieties (Kaye, 1968) only lists just over 200. Most of these are still in existence although many are confined to just a few plants, and hence a cause for concern today. On the credit side many other unnamed cultivars are in specialist collections, perhaps making the total number of different forms now in cultivation something like 500 – or approximately 20% of the total number once grown.

The aims of the National Council for the Conservation of Plants and Gardens (NCCPG) and the BPS, are, of course, not only to conserve these survivors but also to track down old varieties wherever possible. This can be a particularly rewarding pastime. To refind just one old choice form is a thrill and the more elusive these treasures the greater the satisfaction and success. In 1985 (Rickard, 1985) I reported the 'rediscovery' of probable plants of *Polystichum setiferum* 'Plumoso-divisilobum Deltoideum' and 'Plumoso-divisilobum Baldwinii'. I am very pleased to report that these varieties have now been distributed among as many enthusiasts as possible. Other finds are, however, constantly being made. Recently I heard from Peter Boyd, a member of the BPS and Museums Officer of the North Devon Museums Service, how he discovered a collection of ferns growing in the Essex garden of the late F. W. Thorrington – a former committee member of the BPS. Many of these plants, as yet unnamed, now thrive in Peter's Devon garden.

Last year I had an amazing stroke of luck. While on holiday in France I was shown *Asplenium trichomanes* subspecies *pachyrachis*. Although this was a fern unrecorded in Britain, it was immediately recognisable as the form of *Asplenium trichomanes* which grows in the Wye Valley at Symonds Yat in Herefordshire. Scanning Lowe's book *British Ferns and Where Found* (Lowe, 1890), it soon became apparent that this fern was named in the last century as var. *subequale*. It was found by Enys in 1855, also in the Wye Valley. Knowing that the Kew herbarium holds the Thomas Moore herbarium which contains some of Lowe's material, I arranged to check it for the original Enys specimen. This was easily found and it agreed with my Wye Valley fronds. This form turns out to be common in the region and within its range of variation at least one other nineteenth century variety occurs, e.g. 'Imbricatum'. More exciting was the existence of other herbarium specimens of very strikingly different varieties of *Asplenium trichomanes*. These included 'Harovii' and 'Trogyense', recorded from castles in southern Gwent. Hot foot I was off to Chepstow and Caldecot Castles and, believe it or not, these varieties, or at least 'Harovii', were common on both ruins, with particularly fine material on Caldecot Castle. I have collected spores and hope that these will develop into true forms of the varieties to enable plants to be distributed back into cultivation.

Herbaria are, in fact, an overlooked source of valuable information. In the herbarium of the National Museum of Wales I came across a frond of *Polypodium australe* 'Cambricum' collected in 1933. Without any real conviction that it could possibly still be there I eventually tracked down the site to find this beautiful and rare variety adorning both sides of a short drive!

Explorations such as the above can slowly add to our existing range of ferns in cultivation; however, progess is inevitably slow. New additions are more readily obtained by propagation from spores and sometimes old varieties can be re-raised. For example, two very choice varieties – *Athyrium filix-femina* 'Kalothrix', long extinct, was recently re-raised by Ray Coughlin, and *Polystichum setiferum* 'Gracillimum', always very rare, was re-raised by Cor van de Moesdijk, a Dutch member of the BPS. As a result examples of both varieties are now more widely distributed in gardens, although both will probably always be rare.

Given the current situation and the failure of National Collections set up in the past we hope that the present strategy for allocating collections, agreed jointly by the NCCPG and BPS, will be more successful. With the NCCPG setup as an overall watchdog, early signs of difficulties at any of our sites should be spotted in good time.

As a rule we have decided that each fern genus, or part genus, should usually be treated as a separate collection. In each case the aim is that there should be a site where the ferns can be easily seen – the Public Collection – as well as one or two Back-up or Private collections. It is in these collections that the choicer varieties are more likely to be found. In some cases it has been decided to establish Back-up Collections in the North and the South – hopefully there will be safety in numbers – and also collections should be more readily accessible. It is anticipated that the National Collection holders will exchange plants to help build up their collections.

Collections have been sited either in public gardens with a good track record with ferns or in the gardens of enthusiasts where the particular group thrives. We have attempted to correlate the situation of collections with climate and geology, considerations obviously not thought to be of paramount importance in the past when Kew was chosen as the site of the first National Collection of Ferns.

References

ANON., 1895. *Handlist of ferns and fern allies cultivated in The Royal Gardens.* London: HMSO.
CRANFIELD, W. B. 1910. The Jones and Fox collection in the Clifton Zoological Garden, *Br. Fern Gaz.* **I**: 65.
DRUERY, C. T. 1915. Mr W. B. Cranfield's collection, *Br. Fern Gaz.* **3**: 10.
DYCE, J. W. 1989. The Jones nature prints, *Pteridologist* **1**: 262-265.
JONES, A. M. 1876-80. *Nature printed impressions of the varieties of the British species of ferns.*
KAYE, R. 1968. *Hardy ferns.* London: Faber.
LOWE, E. J. 1890. *British ferns and where found.* London: Swan Sonnenschein & Co.
---- 1895. *Fern growing.* London: Nimmo.
RICKARD, M. H. 1985. On the track of old fern varieties, *NCCPG Newsletter.* **1**: 7-6.
STANSFIELD, F.W. 1925. The Kew collection of British ferns. *Br. Fern Gaz.* **5**: 123.

Martin Rickard's initial interest in the distribution of native British ferns soon developed into collecting cultivars of them. His very fine garden houses the national collections of Polypodium, Cystopteris, Oreopteris, Thelypteris *and* Gymnocarpium, *as well as large numbers of hardy and half-hardy exotic species. He is the first Editor of the* Pteridologist, *the Society's journal for articles of wider interest than the academic ones of the* Gazette.

One Hundred Years of Illustrated Fern Books (and then some); A Personal Review

Peter G. Barnes
Royal Horticultural Society's Garden, Wisley, Woking, Surrey GU23 6QB

For many who get the fern 'bug', it can be difficult to say for sure which came first – the interest in growing the plants, or the fascination with the extensive literature on them. For myself, it was neck and neck, Reg Kaye's *Hardy Ferns* being spotted in the local library at about the same time as my first, not too successful efforts to raise plants from spores. It was a serendipitous chance that put that book before my eyes, as I subsequently worked with Reg for a year – most definitely a formative year that set me on my horticultural course.

During that year, I located a second-hand book dealer in North Wales who supplied several of the commoner nineteenth century fern books and thus broadened my interest. Another valuable source was the bookshop in Kendal run by Edna Rock, who quickly came to realise that she could sell me almost anything on ferns, at prices that seem, now, very reasonable. At the time, though, my wage was such that fern books were almost my only non-essential expenditure! There is nothing like a little hardship to reinforce an interest and I have continued to enjoy good illustrated books, not only on ferns, ever since. The comments here are inevitably an entirely personal assessment and reflect the pleasure that I get from such books, both in my own modest collection and in libraries elsewhere.

A curiosity that must strike all new members of the British Pteridological Society quite early on is the fact that the Society was founded well towards the end of what David Elliston Allen calls the Victorian Fern Craze. I make no apology therefore, for delving back much further than our Society's history in the account which follows.

The earliest illustration I have seen, of the sort of fern variant that interests so many in our Society, is in Gerard's *Historie of Plantes* (1597) in which, on page 976 is '*Phyllitis multifida, Finger harts toong*', a typical ramose variant of hart's tongue fern. His '*Trichomanes foemina, The Female English Maiden haire*' appears to be the sort of slightly branched variant of *Asplenium viride* [= *A. trichomanes-ramosum* Ed.] that is not uncommon, especially in cultivation. In all, Gerard describes about 22 ferns. According to Stearn in *The Art of Botanical Illustration* (Blunt, 1950), Gerard's plates are based on those of Tabernaemontanus, themselves copied from the works of earlier artists.

For me, however, the plates in Leonard Plukenet's *Phytographia* are far more interesting. When I first came to work at Wisley, I spent many of my lunch-times exploring the relative riches of the small staff library. My prime interest was to examine the numerous fern books which I had heard of, but had never seen, and one of the more exciting discoveries was the *Phytographia*. The copy here consists of four parts dated 1691 to 1696, bound together with *Almagesti Botanici Mantissa* (1700) and *Amalthaeum Botanicum* (1705), and with manuscript notes in the later parts by William Sherard.

In all, there are 454 full page plates of roughly A4 size, each plate illustrating several plants and including about 148 ferns. The copper engravings are mostly excellent and detailed and are generally quite recognisable, which is by no means always the case with such early works. The plants depicted are in alphabetical order but of course with pre-Linnean names, so that every time the book is opened it is a voyage of discovery. However, among the plates are several illustrations of ferns, both native species and varieties and exotic ones. Of particular interest is plate 284 which shows '*Filix Stafforidensis elegans, foliorii apicibus multicistis Alm. 151 n. 25*', clearly a neatly crested form of lady fern.

The first book devoted to the ferns of Great Britain was James Bolton's *Filices Britannicae*, which appeared in two parts, in 1785 and 1790. Whilst the plates, drawn and engraved on copper by Bolton himself, are of variable quality, the overall effect of the book is most attractive and it is a work I should love to own! There are 31 plates, the whole *'printed on a fine Royal writing paper'*. The text is well interspersed with records of species in the neighbourhood of Halifax, even for such rarities as *Trichomanes*. Some of the descriptions are, like some of the plates, amateurish – thus, *'the root is a short, mis-shapen lump'*, but the book has great charm as well as historical interest.

The prize for the ferniest title page must go to G. W. Francis's little work, *An Analysis of the British Ferns* (first edition 1837). This is another book for which the obvious epithet is 'charming'. The illustrations consist of (2nd edition, 1842) nine full page plates each composed of several tiny, vignette-like copper engravings. A few aberrant forms are included, such as three varieties of *Asplenium scolopendrium*. There are, in addition, wood engravings in the text showing characteristic details of each genus.

A History of British Ferns and Allied Plants by Edward Newman (1840, subsequent editions 1844, 1854 and 1865) marks the beginning of an interest in abnormal or *'monstrous'* (Newman's term!) forms of native ferns. As many as two such forms of lady fern are noted and one is illustrated, whilst three variants of polypody are shown. The illustrations in the book are wood-engravings from his own drawings, but to my eye they have a refined and fresh appearance quite distinct from the mechanical line of later ones. The vignettes ending each chapter are worthy of note too, varying from a detail of the *Trichomanes* sporangium to a rustic cottage or a market hall. The text too is worth reading, being surprisingly detailed and up-to-date in many respects. This is a fine book with which to begin a fern library and, in an obituary (presumably by Trimen, the editor) in the *Journal of Botany* for 1876, it was said to be *'at the time of its first appearance, very much the best book on the subject . . . an accurate and original treatise'*.

Thomas Moore, although a well-known name in the fern world, had much broader interests in the plant world. He was curator of the Chelsea Physic Garden from 1848 and served as secretary of the Royal Horticultural Society's Floral Committee when it was established in 1859. He was also a prolific author on the subject of ferns and his smaller works must still often provide the starting point of many a private collection of fern books. They must have been printed in quite large quantities, since the *Popular History* for example, is not uncommon in second-hand book shops.

A Popular History of British Ferns (1851 and several later editions) is an attractive small handbook describing the native species and mentioning some of the earlier variations from the normal forms. The text is informative and readable and the illustrations are attractive. They appear to be an example of chromoxylographs – in effect, colour wood-engravings. An outline engraving is printed first and subsequently overprinted with successive blocks of flat colour. As sometimes happened with other early colour printing methods, most of the plates show signs of having been touched up by hand. *The Handbook of British Ferns* followed in 1857, being of the 'pocket companion' class of book. In this the illustrations are simple wood-engravings in the text, with no colour. Many variants are described. *British Ferns and their Allies* (undated but *c.* 1860) is an abridgement of the *Popular History* for beginners. The text includes some wood-engravings and there is a section of good coloured plates similar to those of its predecessor but entirely re-drawn.

However, Moore's most notable book is *The Ferns of Great Britain and Ireland . . . Nature-Printed by H. Bradbury*. This appeared in a folio edition in 1855, to be followed by the more often seen octavo one dated 1859. This must represent the high point of naturalistic fern illustrations as well as a milestone in printing techniques. Whilst not the first attempt to use the technique of printing from an imprint of the fronds, it is widely regarded as the classic instance of its successful realisation. An impression of the frond is made in soft metal, which is then hardened to be used as a printing plate. The detail and accuracy of the technique is predictable but the choice of inks in these books results in a delightful light effect, one of the few ferns books whose illustrations are a joy to leaf through for their own sake.

Hooker is one of the great names in British botany, father and son both becoming pre-eminent in their field, both as directors of Kew and as authors of innumerable floras. In addition, Hooker *père* (Sir William Jackson Hooker) was the author of several fern books. Some were not fully illustrated and so are not considered here. However, *British Ferns* (1861), *Garden Ferns* (1862) and *A Century of Ferns* (1854) are in many ways three of a kind, being very consistent in style and also in sharing the same illustrator – the great Walter Fitch. Fitch was a major contributor to the development of the wood-engraving into the standard technique for technical illustration in the middle years of the nineteenth century. The figures in these books however, are hand-coloured lithographs in very much the same format as Fitch contributed to Curtis's *Botanical Magazine* over a period of many years. Fitch has been criticised for a too mechanical style and certainly there is never a hint of a wobbly or weak line. However, I still find the plates in these books to be both accurate and attractive. In addition, there is a certain romance in the mental image of a small team of colourists toiling away to produce what are, in spite of the printed outlines, plates unique to each copy – truly, no two alike. The effect of such hand-coloured plates is far more lively and fresh than the chromolithographs found in other books of the period.

The last remark inevitably leads me to the bulky works of E. J. Lowe, another of the big names in nineteenth century fern circles. Lowe was, to judge by the portrait in his retrospective *Fern Growing* (1895), a dignified character who perhaps took himself rather seriously. He was also a man of wide interests, member of geographical, zoological, microscopical and astronomical societies among others. Again his books have a rather uniform style, the style of illustration in *Fern Growing* being in no way different from those in *Our Native Ferns* (1865-7). In both books, as in the eight volumes of *Ferns British and Exotic* (1862-6), there are chromolithographs of a uniformly drab and uninteresting appearance. Within the text, Lowe uses many wood-engravings which are likewise of a somewhat flat style. Nevertheless, these books all serve as most useful illustrated references. It is only when one moves on to works by Hooker, Moore or Britten that the possibilities of combining accuracy with beauty in illustrations are realised.

One of the Victorian fern books that appears to be quite hard to come by now is *The Fern Portfolio*, by Francis George Heath, a contemporary of Charles Druery. The third edition (1885) claims to include all the British species of fern, illustrated life size. The artist's '*many months of patient toil*' were well rewarded by the end-result, which is a slim folio volume of chromolithographs. These are well reproduced and are indeed most beautiful and realistic in their delineation, even though the registration of some of the plates is not entirely accurate. Heath's *The Fern Paradise* (4th edition, 1878) was one of several rather similar popular works aimed at, and no doubt succeeding in, encouraging the popular interest in ferns. This attractive book has a variety of illustrations, from the 'Woodbury type' frontispiece, a photograph of a ferny coombe, to wood-engravings of ferny scenes in Devon and of specific ferns. The latter are of interest in being white-line engravings on a black background. The author describes them in some detail: '*fronds laid on white card . . . actual impressions taken . . . the figures thus obtained were photographed on the blocks of the engravers*'. In the main, the result is like most effective and accurate silhouettes, some on each plate being hatched in to some degree to give a sense of depth.

Charles T. Druery was, apart from Thomas Moore, one of the few eminent fernists associated to any great degree with the RHS. Although he was not on any of its committees, he was an occasional contributor to its publications and a regular exhibitor of ferns, before its Floral Committee.

His works are among the best of the great efflux of fern books even though, or perhaps because, he wrote towards the end of the Fern Craze. *Choice British Ferns* appeared in 1888. It is a small book that must have provided a good introduction to ferns and their cultivation, with chapters discussing ferny habitats, fern variation, cultivation and propagation, followed by a review of the British native species and a good selection of their variants. An appendix, reprinted from the Linnean Society's Botanical Journal, describes Druery's discovery of apospory in the lady fern. The illustrations in this book differ from the majority of fern books, there being seven plates with the appearance of scraper-boards – presumably these are white-line wood-engravings. The effect is a little startling at first but they serve their purpose well, providing clear impressions of some of the variants now regarded as cultivars. Similar 'reversed' wood-engravings are found in Heath's *The Fern Paradise* (1878). In addition to these, the text is liberally interspersed with more conventional black-line wood-engravings – the standard method of providing good figures in lower priced books during much of the nineteenth century.

The Book of British Ferns (1901, but undated) sees the fairly new technique of illustration by photogravure reproductions of photographs. Gone are the crisp if sometimes rather mechanical wood-engravings, to be replaced by decidedly flat and fuzzy, albeit recognisable, photographs of even greyness. These are quite copious and serve as adequate but not beautiful illustrations to leaven slightly the long lists of fern cultivars, their discoverers and locations.

Druery's *magnum opus*, and probably the most useful of his books to present day fern enthusiasts, is *British Ferns and their Varieties* (1911, but undated on the title page). In many respects this hefty work assembles the best from all his previous books to form an encyclopaedic reference tome. The range of illustrations is different again. Black-line wood-engravings are back, perhaps in recognition of the fact that that they offered the clearest form of accurate illustration available at the time. In addition there are photogravures – still rather grey and fuzzy by today's standards – and there are 40 coloured plates produced by the chromolithographic process. These have the tired, faded look of those in other fern books such as those by Lowe and are very different to the lively and bold plates of Britten's (1910) *European Ferns*.

In spite of the comments above, I have to admit that the nicest illustration in this book, for me, is the photogravure fig. 1, showing the interior of a cool conservatory. If ever a single image epitomises the better side of the Fern Craze, this must be it, a cool looking mass of ferny greenery that one can almost smell.

The other interesting feature of this book is the set of photogravure reproductions of 96 of Colonel A. M. Jones's nature prints of fern variants. As has been mentioned under Thomas Moore's name above, nature printing provides, by definition, the most accurate form of illustration of a single frond other than modern colour photography. Whilst these reproductions are not over clearly done, yet they do convey a sense of real fern fronds that is so often absent from conventional illustrations, and they represent a valuable archival record.

Britten's *European Ferns* (1910) is one of the few books considered here that appeared within the life-time of this Society. The plates are big and bold and strongly coloured – perhaps not refined, but they do make a strong and positive impression. Part of this comes from the printing technique (?chromoxylography) but largely it is due to the inclusion, as in Thorburn's bird and animal plates, of a hint of the habitat. The whole effect of the book is quite different from that of the mass of fern books of the previous century. Another interesting feature of the book is the author's review of previous publications on ferns, with pithy and just comments on such as Bolton, Moore and Newman.

Modern fern books

The publication in 1968 of Reg Kaye's *Hardy Ferns* must have provided a great impetus to the revival of interest in these plants in this country. Although now out of print, it remains available from the Society and has to be the best book to recommend to anyone with an interest, long-established or newly found, in growing ferns. The main value of the book is in its comprehensive coverage of all aspects of the cultivation of ferns – it is less of a guide to their identification. However, it is well illustrated with good photographs in colour and monochrome, showing numerous choice ferns at their best. The review of cultivars is comprehensive and is illustrated by many line drawings. These are much less beautiful than the plates but they are good representations of their subjects – not at all an easy accomplishment with ferns.

Another good source of colour photographs is Roger Phillips's *Grasses, Ferns, Mosses and Lichens of Great Britain and Ireland* (1980). Again, this has some excellent illustrations but it is let down to some degree by the printing, as a comparison with Jones's book (mentioned below) shows clearly. Unlike some other books by Phillips, it is the studio shots that are generally the more successful, if the less attractive, but it remains a useful overview of the commoner (and some less common) native ferns.

About 30 years ago, Eyre and Spottiswoode brought out the 'Kew Series' of pocketable guides to native and cultivated plants. *British Ferns and Mosses* by Peter Taylor (1960), is a handy and useful guide to the native species, with excellent illustrations by Ann V. Webster. The 12 colour plates of ferns are extremely recognisable whilst the line drawings throughout the text manage to combine accurate diagnostic detail with considerable decorative value – a most attractive book.

Christopher Page has produced two solid tomes in recent years. *The Ferns of Britain and Ireland* (1982) must be as comprehensive a field guide to its subject matter as could be hoped for, being clearly and consistently laid out and full of practical observations. The illustrations are the weakest part, although there are a few excellent line drawings of diagnostic details. The bulk of the figures however, comprises faded looking reproductions of photocopies of fronds. These are useful, certainly, not least because each shows something of the range of variation likely to be found within the species. The immediate effect, however, is washed out and drab, a comment which does not under-rate the overall value of the book at all. Page's *Ferns* (1988) is a mammoth account of all imaginable aspects of the subject, with a strong stress on ecological aspects. It is not an illustrated identification guide to ferns but has a large number of attractive habitat photographs in monochrome.

Foreign books

It is important to remember that, in spite of the long interest in abnormal variants, Britain's fern flora is tiny compared to that of many other countries, and occasionally one comes across some nice foreign books in book shops in this country. Colonel R. H. Beddome wrote several books on Indian ferns, mostly attractively illustrated. His *Handbook to the Ferns of British India* (1883, suppl. 1892) is a useful book, that is occasionally seen on the shelves of second-hand book dealers. It has 300 pages of small wood-engravings which are attractive to leaf through as well as appearing to be accurate, with good details of fructification and venation where required.

H. B. Dobbie's *New Zealand Ferns* (second edition, 1921) is one of the overseas classics and is not so easy to obtain now. It is clearly but unattractively illustrated with rather unclear reproductions of photographs, some showing ferns in their habitats, others being studio shots of detached fronds or details of the sporangia. The text is quite entertainingly interspersed with anecdotes concerning fern hunting and cultivation and the book's value is not undermined greatly by the poor illustrations.

In 1975, Barbara Joe Hoshizaki published her *Fern Grower's Manual*, in many ways based on her numerous and well illustrated articles in the American journal *Baileya* (1970). The book lives up quite well to its title and it has a few attractive colour plates. The remaining figures are monochrome photographs, mostly of details or entire fronds and these too are well reproduced. The line drawings scattered throughout the book are best described as functional, as are the rather coarse silhouettes of fronds.

A more recent publication is David L. Jones's *Encyclopaedia of Ferns* (1987) − another book that wholly lives up to its title. In this there are clear monochrome photographs and good line drawings but the glory of the book, apart from the excellent text, is in the abundant colour photographs. In the main, these are very well reproduced (printed in Hong-Kong − a sign of the times). Undoubtedly it is the combination of illustrative styles, and their quality, that makes this book quite outstanding.

Anyone visiting a large book-store in Japan cannot fail to be impressed by the large quantity of books devoted to the native flora. Most of these have copious colour photographs, generally of outstanding reproductive quality. An exception to this is *Shida* by Hiroshi Ito (1985), a functional and basic handbook illustrated by clear line drawings of details and monochrome photographs of entire fronds. Probably a little better known in the West is *Coloured Illustrations of the Japanese Pteridophyta* by Motoji Tagawa (1959 but many times reprinted). This is but one in a wide ranging series of outstanding books on all aspects of natural history published by Hoikusha and it provides a fairly comprehensive review of Japan's very rich fern flora. The full page colour photographic plates show entire fronds. The book is arranged in systematic order and includes keys, a model of what a field guide can be. Although the text is entirely in Japanese, the illustrations make it a most useful book. A more recent pocket guide from the same publisher is Shigeyuki Mitsuta's *Shida no Zukan* (1986). This combines excellent coloured photographs of ferns in the wild, with line drawings illustrating important details. The arrangement is in groups of species with similar *facies*, a convenient idea for a field guide and based on a fairly intelligible illustrated key to these groups.

Finally, one of the most beautifully illustrated photographic fern books of recent years is *Ferns of Malaysia in Colour*, by A. G. Piggott (1988). This has 1,361 colour photographs by C. J. Piggott, beautifully printed and illustrating about 440 species in great detail. The plates include habitat shots as well as close-ups of the fronds.

As indicated at the outset, this is a very subjective and cursory review of an arbitrary selection of illustrated fern books. Hopefully, however, it may serve to show some of those new to ferns an unsuspected aspect of their hobby. Ferns books can be useful tools, or an indulgence; for me they are equally satisfying in either guise.

References
BEDDOME, R.H. 1883, Suppl. 1892. *Handbook to the ferns of British India.* Calcutta: Thacker, Spink.
BLUNT, W.J.W. 1950. *The art of botanical illustration.* London: Collins New Naturalist.
BOLTON, J. 1785-90. *Filices Britannicae.* Vol. 1 Leeds: Binns; Vol. 2 Huddersfield: Brook.
BRITTEN, J. 1910. *European ferns.* London: Cassell, Petter, Galpin.
DOBBIE, H.B. 1921 (2nd ed.). *New Zealand ferns.* Auckland: Whitcombe & Tombs.
DRUERY, C.T. 1888. *Choice British ferns.* London: Upcott Gill.
---- 1901. *The book of British ferns.* London: Country Life & Newnes.
---- 1911. *British ferns and their varieties.* London: Routledge.
FRANCIS, G.W. 1837 (1st ed.). *An analysis of British ferns.* London: Simpkin, Marshall.
GERARD, J. 1597. *The herball, or generalle historie of Plantes.* London: Norton.
HEATH, F.G. 1878 (4th ed.). *The fern paradise.* London: Sampson Lowe et al.
---- 1885 (3rd ed.). *The fern portfolio.* London: Society for Promoting Christian Religion.
HOOKER, W.J. 1854. *A century of ferns.* London: William Pamplin.
---- 1861. *The British ferns.* London: Lovell Reeve.
---- 1862. *Garden ferns.* London: Lovell Reeve.
HOSHIZAKI, B.J. 1970. The fern genus *Adiantum* in cultivation (Polypodiaceae). *Baileya* 17: 97-191.
---- 1975. *Fern grower's manual.* New York: Alfred A. Knopf.
ITO, H. 1985. *Shida.* Tokyo: Hokuryukan Co.
JONES, D.L. 1987. *Encyclopaedia of ferns.* London: British Museum (Natural History).
KAYE, R. 1968. *Hardy ferns.* London: Faber & Faber.
LOWE, E.J. 1862-6. *Ferns British and exotic.* London: Groombridge.
---- 1865-7. *Our native ferns.* London: Groombridge.
---- 1895. *Fern growing.* London: Nimmo.
MITSUTA, S. 1986. *Shida no Zukan.* Osaka: Hoikusha Co.

MOORE, T. 1855 (1st ed.), *et seq. A popular history of British ferns.* London: Reeve & Benham.
---- 1855 (1st ed.), 1859. *The ferns of Great Britain and Ireland . . . nature-printed by H. Bradbury.* London.
---- 1857. *The handbook of British ferns.* London: Groombridge.
---- c. 1860. *British ferns and their allies.* London: Routledge.
NEWMAN, E. 1840 (1st ed.), 1844, 1854, 1865. *A history of British ferns and allied plants.* London: van Voorst.
PAGE, C.N. 1982. *The ferns of Britain and Ireland.* Cambridge: Cambridge University Press.
---- 1988. *Ferns.* London: Collins New Naturalist.
PHILLIPS, R. 1980. *Grasses, ferns, mosses and lichens of Great Britain and Ireland.* London: Pan.
PIGGOTT, A.G. 1988. *Ferns of Malaysia in colour.* Kuala Lumpur: Tropical Press SDN. BHD.
PLUKENET, L. 1691-2. *Phytographia.* London: author's publication.
TAGAWA, M. 1959 *et seq.. Coloured illustrations of the Japanese Pteridophyta.* Osaka: Hoikusha Co.
TAYLOR, P. 1960. *British ferns and mosses.* London: Eyre & Spottiswoode.
TRIMEN, H. 1876. Botanical news. *J. Bot.* **14**: 223-224.

Peter Barnes has been at the Royal Horticultural Society's garden at Wisley for some years, first on the advisory staff and now as a botanist. Trained by Reg Kaye at Silverdale and Jack Drake at Inshriach, he is an experienced horticulturalist whose interests lie in the cultivation of hardy ferns, Hebe, *the Japanese flora and horticultural taxonomy.*

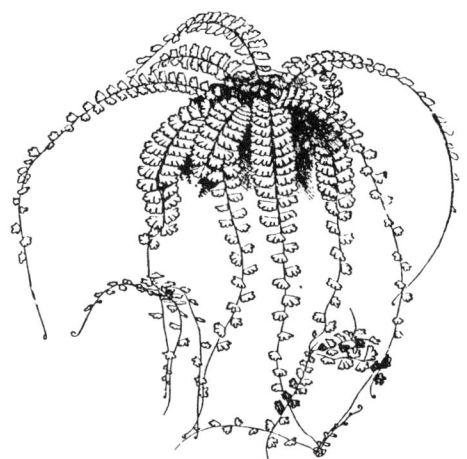

ADIANTUM CILIATUM.

ADIANTUM CAPILLUS VENERIS DAPHNITES.

ADIANTUM DECORUM.

The British Pteridological Society – The First Hundred Years

J. W. Dyce

46 Sedley Rise, Loughton, Essex IG10 1LT

About the middle of the last century a major interest was created by the discovery that our native ferns were producing, in the wild, variations which made desirable garden plants. Books were written in vast numbers and a society, the British Pteridological Society, but NOT our Society, was founded in 1871. Very little is known about this society apart from an *Occasional Paper No.* **1**, published in 1875, which shows that the officers and members included most of the well-known fern hunters and collectors of the day, and a reference in the letter-press to the *Jones' Nature Prints* which appear to have been *begun* by this society, and to the Society's *last* committee meeting held on 26 January 1876. The *Occasional Paper* promised others to follow but, alas! from then on the society disappeared from our ken and we have been unable to find out anything more about it.

In 1891 the Lake District fern collectors banded together to form the Northern British Pteridological Society, with an annual subscription of five shillings (25 pence), and one official meeting annually, to be held on the August Bank Holiday weekend at some local centre, to transact their business affairs, hunt for ferns and hear a fern talk from one of their number. The first meeting was held on Wednesday, 23rd September 1891 at the home of J. Wiper in Strickland Gate, Kendal in Cumbria 'for the purpose of forming an association of fern growers and enthusiastic admirers of ferns for the north of England'. The meeting was conducted in a very efficient manner, chaired by J. A. Wilson. I possess a copy of their Rules, printed in the same year.

Most of the Lake District fern men attended the meeting at which a committee was elected, along with a Secretary, George Whitwell, and a Treasurer, Joseph Wiper. It was decided to invite Dr F. W. Stansfield, a well-known fern man from the south, to become the first President. Two Vice-presidents were elected, Thomas Bolton and J. A. Wilson. As well as being Honorary Secretary of the fledgling Society, George Whitwell found the time to enlarge the third, and last, edition of W. J. Linton's classic little book *Ferns of the English Lake District* in 1893.

The news of the new Society soon spread and attracted members from afar. Among them was Charles T. Druery who was one of the two outstanding fern authorities of the time, along with the President Elect, Dr F. W. Stansfield. They were destined to be the Society's 'power houses' for the next 44 years.

By the time of the second annual meeting in 1892 it was recognised that the 'Northern' in the Society's name made it sound too restrictive and could limit its appeal, especially as it was the first fern society in the field in Britain. Accordingly, the 'Northern' was dropped and the Society became the British Pteridological Society, the name which it continues to enjoy today, in spite of unsuccessful attempts, made on four occasions many years later, to have the name 'Fern' substituted for 'Pteridological'. Members were coming in from all over the country, including survivors from the ill-fated 1871/76 society – among them was G. B. Wollaston, another very prominent fern man from the south. The Society now had 40 members.

The practice of holding one official meeting annually in some ferny part of the country prevailed until the beginning of the Second World War in 1939. The business affairs of the Society were transacted, fern hunts were organised, and talks were given by the more gifted members. From the year 1894 up to 1905 these meetings and talks were fully recorded and published in the Annual Reports sent to members. They contain much very valuable fern information and lore, contributed by the foremost fern experts of the time. Among them were C. T. Druery and Dr F. W. Stansfield. These Reports are valued possessions of the Society which holds two sets, the only two known to still exist, apart from a bound set which I possess, originally the property of Dr Stansfield and inherited by me from his son-in-law, Percy Greenfield, my mentor in all matters relating to ferns. The Society will have the Reports reprinted in book form for sale to members as one of our contributions to mark our Centenary Celebrations in 1991.

Those early years of our Society must have been exciting ones for its members. The 'Victorian Fern Craze' which began in the 1840s, was now, more or less, in its final stages, but its accumulated wealth of fern variety riches was still there in the large and comprehensive collections of the time, whose owners seem always to have been willing and happy to share their treasures with others. They fully recognised the truth of the rather trite saying – the best way to keep a fern (or any plant) is to give it away. In the unfortunate event of a valued fern dying it could always come back to them in the shape of a division or offset from a fellow-grower with whom the original plant was shared. Fern hunting was still actively pursued in the ferny parts of the country which continued to reward the hunters' efforts with a stream of first-rate varieties to add to what can be called, in modern parlance, the 'fern variety lake'. And among the keenest and most successful of those hunters was C. T. Druery.

Druery, in addition to his prowess as a fern hunter, was a most gifted writer on fern matters and, from the 1880s until his death in 1916, contributed a vast volume of printed matter on the subject to all the gardening journals of the time, as well as to the daily newspapers and many magazines. By great good fortune he kept records of all he wrote and our Society possesses eight huge volumes of his press cuttings, assiduously preserved from the time of his earliest writings up to his death. They are veritable gold mines of fern information. He was a very observant man and was the first to discover the phenomenon of *apospory* in ferns whereby the spore stage in the life cycle is bypassed and prothalli develop direct from the sites of the spore heaps. In 1884 he was invited to lecture to the Linnean Society on this discovery which opened a new field of investigation for botanists.

The publication of the Annual Reports ceased with the 1905 one. We do not know the reason. There is nothing in the minutes of the Annual Meetings in 1906 to 1908 which explains why they were not published in those years. The Meetings were held as usual and it may have been because finances were low. But another reason, a more likely one, may have been that Druery had promised better things to come. He had not been idle; already he had had two fern books published, *Choice British Ferns* published by Upcott Gill, London in 1888 and *The Book of British Ferns* in 1901 published by Country Life and Newnes, London. He was the President of our Society in 1901 and the latter book was compiled and edited by him on behalf of the Society, assisted by a committee comprising Dr F. W. Stansfield, J. Edwards, W. Forster, W. H. Phillips, J. J. Smithies, W. Troughton, G. Whitwell and J. A. Wilson, all of them well-known fern men of the time. This book can be seen as an effort to raise the standards of variety collecting and growing. The Preface states *'As ... many of them ... are more curious than beautiful, the need has arisen for a list on purely selective lines, embracing only the pick ... Great pains have been taken by all concerned to confine the list, as far as possible, to really fine symmetrical and constant forms'*. It is good to see them clamping down on the very low standards which had prevailed during the years of the 'Victorian Fern Craze'

Druery was just getting into his stride – his third book, *British Ferns and their Varieties* (one of our Society's two 'bibles' – the other is Reginald Kaye's *Hardy Ferns*) was published by Routledge in 1909. And, having got this off his hands, this prolific writer seems to have been casting around for further outlets to relieve the pressure on his restless creative brain. In the same year (1909) the *British Fern Gazette* was launched under his able editorship. In fact, he literally carried this journal, publishing four issues annually, most of the articles written by himself; he had a phenomenal knowledge of ferns and fern variation.

With the publication of the *British Fern Gazette* the Society, almost overnight, quadrupled its size, new members pouring in from all parts of the British Isles, also from the USA, Canada, Rhodesia (now Zimbabwe), France and Austria. From 1910 until his death in 1917 Druery was the Secretary of the Society as well as the *Gazette* Editor, and the Society flourished in his capable hands.

The death of Druery and the First World War, then reaching a crucial stage, led to a suspension of the Society's activities, except for the publication of the *Gazette* which continued to appear quarterly under the editorship of Dr Stansfield who, with a vast knowledge of fern variation, was, like Druery, able to keep the standards high. Along with W. B. Cranfield he kept the Society alive and when activities recommenced in 1920 the membership quickly built up again to about 100. The subscription was raised to ten shillings (fifty pence) but in 1922 the economic effects of the war made it necessary to reduce the *Gazette* issues to two annually. Although Dr Stansfield was also a prolific writer on fern matters, other members were now contributing more and the journal became less of a one-man effort. The Doctor assumed the additional office of Secretary in 1926 and, like his illustrious predecessor, carried both offices in his stride until his death in 1937.

Some of us today regret very much that Stansfield did not commit his fern knowledge to paper in a book, but he probably felt that Druery had already said it all. This, of course, is not true. In my opinion, he wrote in a more reasoned and balanced way than Druery who was inclined to rush into things with too much enthusiasm and, at times, without due thought. We have to admit that his books have many errors which he should have noticed and it must be concluded that he was always so busy writing that he had no time to spare for the necessary proofing which had to be left to others. A book by Stansfield would have made a very valuable complement to Druery – and a corrective!

With Dr Stansfield's death another crisis was precipitated. W. B. Cranfield, the President since 1920, assumed more active control, while A. H. G. Alston, Assistant Keeper of Botany and in charge of the Fern Herbarium at the British Museum (Natural History) in London, became Editor of the *British Fern Gazette*, and P. Greenfield, Dr Stansfield's son-in-law, was appointed Secretary. With a botanist as Editor and a lack of enthusiastic fern-variety writers, together with the decreasing popularity in ferns, the Society, still very much a horticultural one with a minimal interest in the botany of ferns and relying wholly on its journal to keep in touch with its very scattered membership at home and abroad, entered into a decline. It reached a very low ebb before the beginning of the Second World War in 1939 and during it closed down completely. However, with the December 1939 *Gazette* unlikely to appear, Cranfield compiled a Brochure as a substitute publication. This was a slim volume of eight pages, financed by himself, or possibly with some support from his near neighbours, the specialist fern nursery, Perry's Hardy Plant Farm. Thereafter, nothing more was heard of the Society for the next eight years.

I joined the Society in 1935 with a very keen interest in ferns but it was not until 1939, just before the war started, that I was able to attend what was to be the last of the old style Annual Meetings – the Annual General Meeting, followed by some days in the field fern hunting – which had continued as an established institution since the first meeting in 1891. That meeting made me even *more* keen on ferns!

In 1947, after eight years during which a lot happened in the world – and to me!, – I was back in civilian life again, expecting to hear at any time that the Society was coming back to life. In the end I wrote to the Secretary, P. Greenfield, asking for information. He came up to London to see me in my office and it transpired that nothing was being done as the remaining officers were now old men or were dead and the feeling was that, since the Society was only just existing prior to the war, it should just be left to fade quietly away – just like the 1871/76 Society! This did not suit me at all. I had become a 'fern maniac' in the mid-1930s and attendance at my first field meeting in 1939 had intensified that mania! It was still as strong in 1947, and so far

I had had little opportunity to learn much about ferns. I needed the British Pteridological Society to help me to increase that knowledge, so, at my urging, Greenfield agreed to call a meeting of the Committee – or what was left of it. Six members turned up: A. H. G. Alston, W. B. Cranfield, J. W. Dyce, Rev. E. A. Elliot, P. Greenfield and Professor F. E. Weiss, all of them, except me, very much in favour of letting the Society slip into oblivion. I protested very strongly against this and Greenfield rallied to my side. Between us we succeeded in getting a very reluctant agreement to carry on. For my sins, I was made Treasurer – in the circumstances, an obvious decision since my job was in banking – and as this included Membership Secretary, I had the arduous task of rounding up the membership which had been 'on the loose' for eight years. This was a job after my own heart and I entered into it with great enthusiasm, with the result that within a year I was able to report a paid-up membership of about 100 members. The Society was in business again.

The sad state of affairs before and just after the war was not helped by Cranfield, the President. He had monopolised the office since 1920 and was 80 years old in 1939. A younger man could have put more life into the Society and helped it to come through the war years in better shape. A strong-minded and acquisitive man, very difficult to gainsay, he died in 1948, 88 years old, and still President! He did not seem to appreciate that he was long past his usefulness to the Society and several of his actions in his final years gave me a lot of headaches and extra work later. He was pre-deceased by all the well-known collectors of the past and, as they died, their large collections of the finest fern varieties in existence, gathered together during their long lives, (fern men seem to have a habit of living to good ripe old ages!) were bought up by Cranfield. At the time we applauded this – the ferns were being saved for posterity. With a very large garden and plenty of money, he amassed in this way a vast collection of the cream of fern variation, including a great number of unique varieties.

Cranfield did not believe in giving many good things away and made the most inadequate provisions for the disposal of his huge collection on his death. The result was that what can be called merely a token collection was presented to the Royal Horticultural Society Gardens at Wisley in Surrey to occupy a small open area in the woodland and no attempt was made to ensure that the balance of over 99% of the ferns would be saved by arranging for members of our Society and other people interested in ferns to acquire them. *None* of the well-known old varieties went to Wisley and what they did take was mostly not first-class, and could be compared with taking a bucket of water from the sea. It was the time of food shortages after the war, Cranfield's large garden was acquired by vegetable growers and that huge priceless fern collection, covering well over an acre of ground, which I had seen on my one and only visit to its owner, was *flame-gunned*! The cream of over 100 years of accumulation of first-rate fern variation went up in flames!

With all the big collections finishing up in Cranfield's garden, very few of the ferns were in other hands, and today we are doing our best to trace them and gradually replace a fraction of the loss. It was quite some time after his death that we in the Society learned what had happened. The new owners, acting in a much more responsible manner, invited the people in the surrounding area to come and help themselves to any plants before they were destroyed. Unfortunately, none were fern collectors or members of our Society. It was this situation in our Society that determined me in later years to ensure that presidents should be elected for only a limited tenure in the office. It was Dr Holttum's decision when he resigned after three years as President, insisting that this period was long enough, that enabled the committee to establish this practice. Later, a tradition was established so that the office is held alternately by botanists and horticulturalists, in recognition of the Society's interests.

Fortunately, Cranfield's inadequate arrangements for the survival of his ferns were an exception. In more recent times collections have been passed on successfully to

Reginald Kaye's nursery at Silverdale in Lancashire, to Fibrex Nurseries at Pebworth in Warwickshire and to Mrs J. K. Marston at Nafferton in Yorkshire. The Bolton collection, built up over the years since the time of Thomas Bolton, a founder member of the Society in 1891, has been greatly depleted by the generous passing on of good varieties to other collectors. At present I am passing on my collection to Martin Rickard and making sure, while I am still alive, (I am now approaching my 86th Birthday) that the right ferns – the valuable varieties – are safe in good hands. Also, it is very satisfactory to add, the Society is at present working in conjunction with the National Council for the Conservation of Plants and Gardens (NCCPG) to set up national collections of plants to ensure the preservation of our best varieties. Several of our members are participating in the scheme.

The late 1940s and the next decade in the 1950s was one of change. The membership remained static at about 100, but not the officers, with one exception – the Treasurer! Because of this I found myself carrying more and more responsibility for the running of our affairs. After Cranfield's death in 1948 Robert Bolton held the Presidency for one year before he too died, followed by A. H. G. Alston for nine years till 1958. On his death Thomas H. Bolton held the office for two years then resigned in 1960. P. Greenfield resigned as Secretary in 1947, having held the office since 1937 – he was no longer a young man and found the post too onerous. He was succeeded by J. R. Pulham, a protégé of Cranfield's, for two years. He was not a fern man nor a member of the Society and was, we felt, browbeaten into the office by the strong-minded President! When Alston became President in 1949, the Rev. E. A. Elliot took over the editorship of the *Gazette*, and added to this the Secretaryship in 1951, holding both offices until his death in 1959.

Sorting out this 'confusion', it will be seen that three officers, Alston, Elliot and Dyce, controlled the Society for most of the 1950s. But, Alston took little part in our routine work and Elliot, as well as holding two offices, was struggling nobly with a terminal illness which he used to dismiss lightly as his 's.o.b.' – his 'spot of bother'. Accordingly, both his jobs as Secretary and Editor suffered and he leaned heavily on me. Fortunately, in the background was Greenfield, a tower of strength, who was not only my mentor in fern matters but my adviser in running the Society. It was all good training for me and when I was elected Secretary in 1959 on the death of Elliot I already had the job at my fingertips. At the same time, Greenfield and I decided to invite A. Clive Jermy, a young botanist who had succeeded Alston as Head of the Fern Section at the British Museum (Natural History), to become editor of the *British Fern Gazette*.

Being very much tied to my work in the City of London – it was before the days of 'worker emancipation' and generous holiday leaves! – I did not have too much time to devote to the Society, but Elliot, Greenfield and I did manage to inaugurate a fuller meetings programme during the 1950s. The West Country, Devon and Somerset mostly, was our chief attraction and the three of us used to spend a week annually fern hunting in the lanes. Over the following years we were joined by more and more members and the annual field excursion was revived. About this time a Society friend who had gone to live in Wales near Cardiff invited me to spend a ferny weekend with him. In the following year when the invitation was renewed, I spread the news among a few members likely to be interested. This proved very successful and so our weekend field meetings were started. Before the decade finished I had a good annual programme of weekend meetings in operation.

In the 1960s a new era began. Both Clive Jermy and I were vigorous in our respective fields and the Society, which had been more or less slumbering since the war, now began to come to life once more. Dr Eric Holttum was elected President, a very auspicious start to the new regime. He was one of the world's foremost and most distinguished botanists who spent most of his working life in Malaysia in charge of the Singapore Botanic Garden and later as Professor of Botany at the University of Singapore. Ferns were one of his chief interests and we are very proud and honoured to have had him as a member of our Society and as our President. Some years later another of our distinguished members, Professor Irene Manton, who occupied the Chair of Botany at Leeds University and was in the forefront of world fern research, honoured us by accepting the office. A measure of her standing in the botanical world was her Fellowship of the Royal Society and her appointment to the Presidency of the Linnean Society of London.

Among our non-botanist Presidents, Reginald Kaye stands out. He has spent a lifetime among ferns and is the owner of Britain's largest and most comprehensive fern nursery at Silverdale in Lancashire. In more recent times, when I retired from the Secretaryship, I was given this ultimate honour of President, followed by a further honour, on the termination of my period in the office, when I was elected our first President Emeritus. Clive Jermy, with whom I have worked in close collaboration for so many years, followed me in the Presidency. The present holder of the office is Dr Barry A. Thomas, Keeper of Botany at the National Museum of Wales in Cardiff. His chief task will be to see the Society safely and successfully through our Centenary celebrations in 1991. I continued to hold both offices of Secretary and Treasurer until 1974 when Barry Thomas, then at the University of London, Goldsmiths' College, relieved me of the latter. Five years later I resigned as Secretary and was succeeded by A. R. (Matt) Busby who still holds the office.

My first five years as Secretary were rather harassing ones. Our new Editor was a botanist, with the result that botany began to feature more and more in the *British Fern Gazette*. This was a natural development as we simply could not persuade our fern growers to provide material for publication. There were, however, plenty of botanists eager to take advantage of this new outlet to get their research papers published. The *Gazette* became more and more botanical, and the non-botanical membership maintained quite rightly that they did not pay their subscriptions to receive a botanical journal. I did my best to encourage them to give us material to publish in the *Gazette*. One or two did and I began to write more and more, but I was still very much feeling my way and broadening my knowledge of fern variation.

Meanwhile, in Clive Jermy's capable and enthusiastic hands (helped considerably by Jim A. Crabbe, also on the Museum staff) the *Gazette* was going from strength to strength and soon became one of the world's leading botanical journals. In 1974 the 'British' was dropped and it became the *Fern Gazette*. Botanists, botanical institutions and universities worldwide began to join the Society as subscribers in increasing numbers. Dr C. N. Page followed Clive Jermy as Editor-in-chief for seven years until 1983, since when the Editorship has been a joint effort with permutations of the following botanists – Jim A. Crabbe, Dr Mary Gibby, Dr Chris N. Page and Dr Barbara S. Parris. At the time of writing Jim Crabbe is holding the office on his own.

When I retired from banking at the end of 1966, glad to shake the dust of the City of London off my feet, I was able to make further contributions to the Society. With plenty of time on my hands I decided I could now do something for our horticultural members. One of my secretarial duties was to send out an annual news-sheet to the membership, consisting of one, sometimes two, foolscap sheets run off on a duplicating machine. This was before the days of cheap photocopying. In 1967 I expanded this to 26 pages; I gave full reports on all the meetings of the year, a detailed programme for the following year and wrote an article on fern holidays in Britain with information on various good fern centres. An article by Druery on fern culture, which had been written for *The Garden* magazine in 1901, was included and I persuaded one of our photographic members, P. R. Chapman, to contribute an article on photographing ferns. I finished with a three-and-a-half page selection of fern literature.

This achieved the desired result. Our horticulturalists ceased to complain and instead I received articles to publish! We now had a second journal, a very amateur effort, 'tis true, but it was a beginning. With each annual issue I become more proficient and at the end of six years the *Newsletter* developed into a 'real' journal, the *Bulletin*. And here I have to pay tribute to an ally who ably supported me and made it all possible – Valerie Metcalfe of Metloc Printers Ltd in Loughton, Essex. She has continued to print the Society's publications ever since.

Events now began to move more rapidly. With two journals published annually, the membership began to increase in numbers. Botanical and scientific institution subscribers increased and research botanists became more aware of our *Gazette* as an outlet for the publishing of their pteridological papers. Its world reputation increased accordingly. Our gardeners and fern collectors began to contribute more to the *Bulletin* which, by the end of the first volume of six parts, was successfully fulfilling its role. With Volume 2, Martin Rickard took over the Editorship from me. As a dedicated collector of our British fern varieties he is gradually building up a large and important collection. Later, when we launched our third annual journal in 1984, the *Pteridologist*, aimed specifically at the grower and collector, he became its first Editor, and the *Bulletin* was left to perform the role of a house journal, with the Secretary, Matt Busby, as its Editor. Its function now is to cover Society business and to report our various activities in more detail. Alison M. Paul took over as Editor in 1990.

In the same year that the *Pteridologist* was launched we began to publish small books on fern matters. The first of these *Special Publications* was *A Guide to Hardy Ferns* by our member Richard Rush and has proved to be very popular with our members worldwide. *Special Publication* No. **2**, *Fern Names and their Meanings* by J. W. Dyce, was published in 1988. This also is selling well worldwide, although its interest is clearly more specialised. A third *Special Publication* is scheduled for publication during our centenary year, *The Cultivation and Propagation of British Ferns* by J. W. Dyce [published May 1991. Ed.]. We are also planning two other books to celebrate the Centenary. One will be a 'coffee-table' book entitled *'A World of Ferns'* which should have wide general appeal, written by Josephine M. Camus, Clive Jermy and Barry Thomas [published June 1991. Ed.]. A collection of articles written by Society members will be presented as *The History of British Pteridology 1891-1991*, to reflect the history of the Society and developments in various aspects of the diverse interests in ferns of its members [i.e. this volume. Ed.].

From the 1960s the annual meetings programme developed further with much valuable assistance from R. F. (Dick) Cartwright, another retired member who lived near me. Later, the job was taken over by A. J. (Tony) Worland as our first Meetings Secretary. As the years passed other members became actively involved and added to our national programme by becoming regional secretaries with their own local programmes. Currently, they are – Paul H. Ripley, South-East England; Martin Cragg-Barber, Wessex; Matt Busby, West Midlands; Jack H. Bouckley, Leeds and District; Bernard Williams, East Anglia. Our present Meetings Secretary is Anthony C. Pigott who is also in charge of promoting Society merchandise.

Field meetings give plenty of opportunity for the recording of fern distribution throughout the country. I began this when, as Secretary, I made a point of attending all field meetings. We appointed Tony Worland as our official Recorder, to collate the records sent to him and pass them on to the national recording scheme, run by the Biological Records Centre at Monk's Wood, Cambridgeshire. Our work added much to make the *Atlas of Ferns of the British Isles* more complete. The *Atlas* was a joint publication in 1978 by the Botanical Society of the British Isles and ourselves.

A very worthwhile activity is our Spore Exchange which was initiated by David Russell and which has been developed over the years by Dick Cartwright. Thanks to his devoted work it is now well supplied annually with spore contributions from home and abroad, and a long list of available spores in the Exchange is sent out annually. There is a worldwide demand for them and the Organiser is kept busy. Richard Cartwright retired from the job in 1988 and the present Organiser is Margaret Nimmo-Smith who is enthusiastically carrying on the good work. In 1984 Richard Lamb took over the Plant Exchange, also initiated by David Russell, and this much appreciated facility is now in the hands of Rosemary Hibbs.

Another service is our Reading Circle, started many years ago to circulate the *American Fern Journal* among members. This is at present organised by our Secretary, Matt Busby.

In an early number of the *Newsletter* I published an article on old fern books and intimated that I had acquired a few unwanted ones which I was prepared to sell for Society funds. They sold quickly and I found there was a demand for old and new fern books by our members. Thus began BPS Booksales. I broadcast to the membership that I was prepared to buy, at a reasonable cost, unwanted books on ferns, and I spent a lot of time, wherever I went, browsing in old second-hand bookshops. The next step was to approach publishers of new fern books, asking if they would be prepared to concede trade discounts to me on purchases made from them. Without exception, I have always had a favourable response.

The lists now contain 60 or more titles annually. As I write, in 1990, I have decided, because of my age – but, even so, very reluctantly! – that the time has come to pass on to some younger and more energetic member this last of the many offices I have held in the Society. This job of buying and selling fern books is one which has been most satisfying and has given me much pleasure over many years and I hope that whoever takes it on from me will find it an equally pleasurable task. [Steve J. Munyard, another avid bibliophile, succeeded Jimmy Dyce in 1991. Ed.]

About the time that BPS Booksales was started, fern interest in Australia commenced to escalate and Australians began to join the Society in increasing numbers. Applications for help were received by me from many people wanting information and books on ferns. BPS Booksales did some very brisk business with Australia during this period and our Australian membership rapidly increased. Now, Australia is very much on its own feet, with three large fern societies – the 'Australian Fern Craze' took off, just as ours did over a century ago, and books on ferns are being published in increasing numbers. I treasure a remark by one of my Australian visitors – 'We in Australia regard your Society as the Mother Society'. That remark alone made me feel that all my work and time spent in helping Australia to find her 'fern feet' had been very much worthwhile.

As the membership increased the work of Treasurer combined with Membership Secretary became more demanding and in 1977 Barry Thomas passed on to Lt. Col. Phillip G. Coke the job of Membership Secretary. Later, in 1988, when he became President, after serving 14 years as Treasurer, Barry Thomas passed this office on to Dr Nick J. Hards who continues to hold it. And after seven years as Membership Secretary, Phillip Coke passed on this office to the present holder, Alison M. Paul, another botanist at The Natural History Museum. In 1980 we appointed an Archivist, Nigel A. Hall, with the duty of gathering together and recording events of note in the Society. In the years to come this should become a more and more important activity.

Over the years we have been closely associated with The Natural History Museum, London, where the successive fern taxonomists have lent their able support to the Society. We have also enjoyed close ties with the Royal Botanic Gardens, Kew, for very many years and the present Assistant Curator, John R. Woodhams, whose previous post was Supervisor of the Fern Houses, contributes much to the success of the Society. Mention must also be made to his predecessor in charge of the Fern Houses, the late H. J. (Bert) Bruty, who was one of our strong supporters. His work at Kew was recognised in the award to him of the British Empire Medal, an honour which was well deserved. The Society has organised two international symposia in the last twenty years. The first, on the phylogeny and classification of the Filicopsida was held jointly with the Linnean Society in London, 13-14 April 1972, and the Proceedings were published by Academic Press in 1973 as Supplement No. 1 to the *Botanical Journal of the Linnean Society*. The second was held in Edinburgh, 12-16 September 1983 on the biology of pteridophytes with the Royal Society of Edinburgh and the Linnean Society as co-sponsors with our Society. The papers presented there were published in the *Proceedings of the Royal Society of Edinburgh,* Section B, volume 86 in 1985. A third symposium, on the propagation and culture of pteridophytes, is scheduled for 1991.

Our Society, in spite of some difficult periods, has weathered all the storms and has come safely through to complete its first hundred years. It is growing stronger as the years pass, with a present membership of about 700, comprising members and subscribers, scattered throughout over 40 countries in the world. Interest in ferns continues as strong as ever, and in recent times this has resulted in more and more attention from research botanists. New fern species continue to be found, chiefly in the more inaccessible parts of our globe and the present world total of about 12,000 species is still increasing all the time. I can visualise that in our second century we shall become even more of an international power in the world of ferns, with the increasing interest 'rubbing off' to stimulate a greater activity in the more humble ranks of collectors and growers of species and varieties. I see no reason for doubt that, in due course, the British Pteridological Society will be celebrating its second century of existence, an even more impressive Society than at present.

Jimmy Dyce is renowned as the man who saved the Society from extinction. A simple enquiry from a friend developed into a mania fostered by his mentor, Greenfield. His unswerving devotion to ferns and the Society is outlined in the articles by himself, Busby and Hall in this volume. His chief interests have been the recording of fern distribution in Britain and the growing of many fine varieties of our native species.

Officers of the British Pteridological Society during its first hundred years – 1891-1991

Presidents
1891	The first meeting of the Society was held on the 23rd of September 1891 at Kendal in Cumbria. J. A. Wilson was elected Chairman, rules were formulated and a secretary and treasurer appointed. It was decided to invite Dr F. W. Stansfield to become the first president.
1892-1897	Dr F. W. Stansfield
1898-1901	C. T. Druery
1902-1904	Dr F. W. Stansfield
1904-1907	W. H. Phillips
1907-1908	Dr F. W. Stansfield
1908-1909	J. J. Smithies
1909-1920	A. Cowan (resigned this year)
1920-1948	W. B. Cranfield (died in this year)
1948	Robert Bolton (died in this year)
1949-1958	A. H. G. Alston (died in this year)
1958-1960	T. H. Bolton (resigned this year)
1960-1963	Dr R. E. Holttum
1963-1966	R. Kaye
1966-1969	Dr J. Davidson
1969-1972	Professor I. Manton
1972-1975	H. L. Schollick
1975-1979	Dr S. Walker
1979-1982	J. W. Dyce
1982-1985	A. C. Jermy
1985-1988	Mrs G. Tonge
1988-1991	Dr B. A. Thomas

Secretaries
1891-1909	G. Whitwell
1910-1917	C. T. Druery
1917-1920	W. B. Cranfield
1920-1926	C. Henwood
1926-1937	Dr F. W. Stansfield
1937-1947	P. Greenfield
1948-1950	J. R. Pulham
1951-1959	Rev. E. A. Elliot
1959-1979	J. W. Dyce
1979-	A. R. Busby

Treasurers
1891-1903	J. Wiper
1903-1909	W. Wilson
1910-1920	W. B. Cranfield
1920-1921	J. J. Smithies
1921-1926	C. Henwood
1926	Dr F. W. Stansfield
1926-1938	J. J. Sheldon
1939-?	Dr T. Stansfield (died pre-1947)
1947-1974	J. W. Dyce
1974-1988	Dr B. A. Thomas
1988-	Dr N. J. Hards

Membership Secretaries - removed from Treasureship to become separate office
1977-1984   Lt. Col. P. G. Coke
1985-   Miss A. M. Paul

Editors of *The British Fern Gazette* - later changed to *The Fern Gazette*
1909-1917   C. T. Druery
1917-1937   Dr F. W. Stansfield
1937-1948   A. H. G. Alston
1949-1959   Rev. E. A. Elliot
1959-1965   A. C. Jermy
1966-1968   A. C. Jermy, J. A. Crabbe
1969-1970   A. C. Jermy, J. A. Crabbe, Dr F. M. Jarrett
1971-1972   A. C. Jermy, J. A. Crabbe, Dr B. A. Thomas
1973-1975   A. C. Jermy, J. A. Crabbe, Dr C. N. Page, Dr B. A. Thomas
1976-1978   Dr C. N. Page, J. A. Crabbe, A. C. Jermy
1978-1980   Dr C. N. Page, J. A. Crabbe, J. W. Grimes, A. C. Jermy
1981-1983   Dr C. N. Page, J. A. Crabbe, A. C. Jermy
1984   Dr M. Gibby, Dr C. N. Page, Dr B. S. Parris
1985-1986   Dr M. Gibby, Dr B. S. Parris
1987   J. A. Crabbe, Dr M. Gibby, Dr B. S. Parris
1988   J. A. Crabbe, Dr B. S. Parris
1989-   J. A. Crabbe

Editor of *The Newsletter*
1963-1972   J. W. Dyce

Changed format into *The Bulletin*

Editors of *The Bulletin*
1973-1978   J. W. Dyce
1979-1983   M. H. Rickard
1984-1986   A. R. Busby
1987-1989   A. R. Busby, Miss A. M. Paul
1990-   Miss A. M. Paul

Editor of *The Pteridologist*
1984-   M. H. Rickard

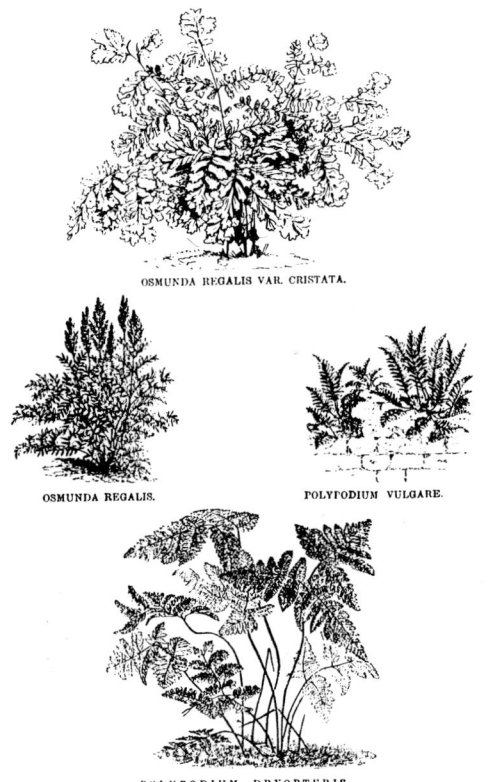

Gleanings from the Minute Book 1891 - 1991

A. R. Busby

'Croziers', 16 Kirby Corner Road, Canley, Coventry CV4 8GD

I have, in my temporary possession, the Society's Minute Book, which was purchased in Kendal, Cumbria during September 1891. It measures 13 by 8.5 inches (33 by 21.5 cm), is 2 inches (5.1 cm) thick and weighs some 6.5 pounds (2.9 kg). As I had to carry it to each annual meeting, I considered it to be the Secretary's curse. Yet, within its battered brown covers lies a treasury of information and facts on the birth and development of our Society.

Page one records the first meeting held at Mr Wiper's home at Stricklandgate, Kendal, on the evening of the 23rd September 1891 at which the rules of the Northern British Pteridological Society were formulated. This meeting was notified by George Whitwell and Robert Whiteside, to *'most of the Fern enthusiasts in Kendal and District'*. The report of the meeting is introduced by the following preamble: *'For years past a feeling had been pregnant in the minds of some few northern pteridologists that a Society or Association should be formed for the purpose of collecting and recording the information acquired by existing fern collectors, for naming any variety or specimen that is to be found or raised, and acquiring any information and recording same in the Society's Minute Book that may appertain to ferns in general; also for making it a meeting place where pteridologists could meet and obtain a broader and larger acquaintance with ferns and fern collectors throughout the north of England'*.

Some sixteen pteridologists were notified of this meeting, most from the Cumberland and Lancashire area but with one or two from Manchester and Sheffield. Those in attendance were G. Whitwell, J. A. Wilson, J. Stewardson, T. Bolton, J. Gott, J. Wiper, J. Garrett, J. J. Smithies, W. Wilson and R. Whiteside. J. A. Wilson was elected Chairman, *'he being the oldest pteridologist in the room'*. R. Whiteside was elected as Secretary for the time being, and it was decided that F. W. Stansfield should be invited to be President. Before the draft rules had been discussed, changed and passed, it was agreed that the Society would be called the Northern British Pteridological Society. One of the last motions passed at this preliminary meeting was that a report of this meeting should be forwarded to *Amateur Gardening* and *'Kendal papers'* respectively.

This first meeting was followed by a Committee meeting held at Mr Wiper's room, Stricklandgate, Kendal on the 25th November 1891. Obviously the reports to the press had paid off for at this meeting C. Druery, W. H. Phillips, J. A. Barnes, George Stabler, P. Neill Fraser, W. B. Boyd, James Wrigley, William Askew, J. G. Newsham, H. L. Storey, J. Bolton and T. Clark, were proposed and passed for membership.

At the next meeting held on the 3rd February 1892, again at Mr Wiper's room, a letter was read from G. B. Wollaston offering to give a talk on ferns at the next AGM on the *Classification and Fructification of Ferns*. Amongst other business, the design and acquisition of membership cards was discussed. A few of these cards still survive. This first AGM was held at Windermere on the 1st August 1892, attracting twenty-four members and a motion was passed agreeing to drop *Northern* from the title. It was also at this meeting that the habit of exhibiting fronds of fern varieties was established and the names entered into the minute book. Exhibits at this first AGM were: *Athyrium filix-femina* 'Todeoides Cristatum Troughton', *A. f-f.* 'Erecto-pinnatum Stewardson', *A. f-f.* 'Cruciatum Cristatum Tyldesley', *Polystichum angulare* (= *P. setiferum*) 'Tripinnatum Phillips', *P. angulare* 'Setoso-concavum Stewardson', *Polystichum aculeatum* 'Alatum Stewardson', *Lastrea filix-mas* (= *Dryopteris filix-mas*) 'Polydactyla Stewardson', *Scolopendrium vulgare* (= *Asplenium scolopendrium*) 'Ramo-marginatum Stewardson', 'Bicornutum Ramosum Stewardson' and 'Marginato-peraferens Stewardson'.

Offers of papers to be read at meetings now featured prominently in the Committee business. Mr Earlson offered a paper on *Hybridisation* and Mr Phillips a paper on *Fern*

Hunting in the Irish Wilds while Mr Stewardson or Mr Druery were to be asked to submit papers.

At the third AGM held at the Phoenix Rooms, Lancaster, a resolution was submitted by Mr Whiteside empowering the Society to grant *'Certificates of Merit in degrees for fern varieties collected or raised from spores and whose character were regular and permanent'*. This might appear to reward the raiding of the countryside for ferns; however, conservation had already come to mind. At the Committee Meeting held at Mr Wilson's house, Bowness, a resolution was to be submitted at the next AGM relating to the way that *'the country had been plundered of its wild flowers and ferns, and its beauty despoiled. If approved, a memorial would be submitted to the County Council.'* Sadly, at the next (fourth) AGM held at Mr Webb's rooms, Stramongate, on 5th August 1895, no mention of it is made. One other matter that was touched on at that meeting was the pleasure expressed by the Chairman (Mr Stewardson in the absence of F. W. Stansfield) on behalf of the members at the presence of several ladies, this being the first occasion when ladies had attended. The ladies are not named but may have been the wives of members present.

The fifth AGM, held at the Institute, Bowness on the 3rd August 1896, heard two papers which reflected the varying approach amongst members to their interest in ferns. Charles Druery read his paper entitled *The Marvellous Side of Fern Life* and this was followed by Dr Stansfield's paper entitled *Weissman's Theory of Heredity and Its Relation to British Ferns*.

By 1900 the Society membership stood at 45, having been steady at 42 for some years. On the 6th August 1900, the Society held its ninth AGM at the Institute, Bowness. The Society made a presentation to G. Whitwell of a *'massive handsome clock, a choice full set service of china and an illuminated testimonial by the members of the Society'*. The Treasurer's report for August 1901 reveals some interesting points. A Mr J. C. Williams purchased his Life Membership for three guineas (£3.15). Hire of room and waiters was only ten shillings (50p). It was policy in those days that anyone wishing to join the Society had to be proposed and seconded by a current member. Occasionally a death was sadly reported, none more so than the death of G. B. Wollaston in 1898. The Minute Book records the letter of condolence sent to Mrs Wollaston by George Whitwell on behalf of the Society.

Throughout the early years of the century, the meetings continued along similar lines with the AGMs usually held in Bowness and always incorporating an exhibition of new fern varieties to which names were given. The minutes of the 15th AGM held on the 6th August 1906 record a slight change. Arrangements were made for a drive to Wythburn. *'The weather being fine, a most enjoyable day was spent though no finds of importance were made.'* Between 1906 and 1910 the membership leapt from 52 to 115. However, no reason for this sudden upsurge is mentioned in the minutes.

During these early years, the Society published a yearly abstract of reports which gave the membership the opportunity to read the papers presented at the annual meeting. This formed the predecessor to the *British Fern Gazette* which was first published in 1909.

The Society has proved nothing if not adaptable. The minutes of the 19th AGM held on the 1st August 1910 record that it was held in the waiting room at Beattock Station, Dumfriesshire with seventeen members in attendance. What the ordinary travelling public made of this is not recorded but it does conjure up an extremely bizarre scene! This meeting was based at Moffat and the following supplementary note is attached to the minutes:- *'Prior to the business meeting, several informal ones were held on the Friday and Saturday at the Annandale Arms Hotel, Moffat, where the President, Mr Alex Cowan, had kindly placed his sitting room at the members disposal. A considerable number of fern fronds were exhibited including some by Mr Druery of the new 'Gracillimum' seedlings of* Polystichum aculeatum *raised from* P. aculeatum *'Pulcherrimum' and some*

Polystichum aculeatum-*like seedlings of* Polystichum lonchitis *raised by Mr Boyd which were considered to be bipinnate sports of* P. lonchitis, *the narrow form of fronds of the species being maintained. 'The several fern-hunting expeditions made were greatly handicapped by stormy, wet and chilly weather and only resulted in the finding of some angustate and crispate forms of* Lastrea montana (= Oreopteris limbosperma) *of full promise but which, owing to the exposed habitat in which they were found, it was commented, could not be safely named until they had been tested for constancy. Hence nothing really new or striking could be recorded.'* The Society had now begun to explore areas further afield and at the 20th AGM (1911) held in Barnstable, the matter of the use of cars during meetings was raised. The President, Mr Alex Cowan, circulated to the members of the Committee, a letter proposing the allocation of a sum of money from Society funds towards the cost of vehicles used for conveying members at meetings and towards the hire of a sitting room that could be used as a members' common room. All approved. The Society's activities were now overshadowed by the approach of the Great War. At the 24th AGM held on the 2nd August 1915 at 11 Shaa Road, Acton, London, home of the Secretary, Charles Druery, a letter was addressed to all the Committee and Officers, at the suggestion of W. B. Cranfield, F. W. Stansfield and C. Druery. Mr Druery's letter read: '*A strong feeling has now been expressed by several members of the Committee that in consequence of the war, the usual autumn excursion should be abandoned this year. It will however, be necessary to hold a formal meeting for the purpose of passing the accounts as audited, and electing officers, auditors etc., who it is hoped, under the existing circumstances will kindly consent to act during the ensuing year. In consequence of my state of health, I am unable to undertake the journey to Kendal and would therefore ask you to approve the accompanying resolutions by signing the same and returning them to me in order that I may issue the requisite notice to members.*' It was approved and a letter was sent to all members explaining the actions of the Society. No further meetings were held and the next entry records the sad loss of C. T. Druery, VMH, FLS, who died on the 8th July 1917. There follows a long obituary by F. W. Stansfield outlining Mr Druery's total involvement in the world of fern growing and his long dedication to the Society.

The Society resumed activities with an AGM on the 2nd August 1920 at Kendal, with Alex Cowan as President, W. B. Cranfield as Secretary and J. J. Smithies as Treasurer. Two notable decisions were made: to increase the subscription to ten shillings (50p) for new members (with an option to members of 15 years' standing to pay the old rate); and to affiliate the Society to the Royal Horticultural Society.

August 1921 found them at the George Hotel, Chard where the President decided that the time was now ripe for the re-issuing of the Certificates of Merit. During these years, the Society's financial situation must have been dire as no accounts are appended although mention is made of their being audited and published in the *British Fern Gazette*. Several members of the Committee occasionally gave donations of several pounds to the Society, which probably enabled the Society to keep its head above water at that time. It is recorded in 1922 that, after all debts had been paid, the Society's sum in hand was £10 10s. (£10.50).

At a Committee Meeting held at W. B. Cranfield's office in Cheapside, London on the 25th June 1926, it was announced that Mr Charles Henwood, Honorary Secretary and Treasurer had found it necessary to file his Petition in Bankruptcy and had sent in his resignation of his offices in the Society. The President, W. B. Cranfield, had obtained the agreement of the official receiver that the Society's funds should be looked on as trust funds and that they would be paid out in full. According to Mr Henwood, the Society's funds stood at £33 1s 1d (£33.07). At the 1926 AGM held at the Bull Hotel, Bridport, this figure had been revised to £42 10s 11d (£42.55). Membership stood at 41 paid members, leaving 35 who had neither resigned nor paid, but in spite of their money worries fern growing continued apace!

At this same meeting, T. B. Blow 'exhibited a remarkable successful culture of Athyrium filix-femina 'Clarissima' raised by apospory and presented the plants raised to the members present'. Then followed a vote of thanks and of cultural commendation to 'Mr Blow coupled with the names of Mr Henwood who supplied the aposporous material and Mr Henry Mount who laid down the frond for culture'.

It was also during these years that the Society was beginning to address itself to the scientific side of pteridology and acknowledged its responsibilities for accurate naming. During the 1922 AGM held at Padarn Villa Hotel, Llanberis, the minutes report that 'a long discussion took place on the question of the revised nomenclature as presented by Dr G. C. Druce in the December issue of the British Fern Gazette', as part of the general revision of this subject at the Vienna Conference. Some doubt was expressed as to whether the settlement as presented by Dr Druce would be a stable one, and eventually it was resolved that a Sub-committee consisting of the President, W. B. Cranfield, and the Editor, F. W. Stansfield, 'be appointed to confer with the Royal Horticultural Society and the Kew Authorities in order that the Society should come into line with the leading scientific bodies on the question'. As we know to our cost, nomenclature in ferns has never proved stable!

Another sacred cow was put up for slaughter during the 33rd AGM held on the 3rd September 1928 at the Digsby Hotel, Sherborne, Dorset. A letter from A. J. MacSelf to the President suggested that the Society should change its name from its 'Greek form to the Anglo-Saxon equivalent of the British Fern Society'. The minutes report: 'Considered discussion took place on the subject of this letter and while the general feeling seemed to be against a change in the naming of the Society, it was recognised that there was something to be said for the other side and it was eventually agreed that the Society should keep its original name but that an explanatory phrase or subtitle 'the British Fern Society' might upon occasion be used in brackets after the name'. Little did they realise that this matter would re-emerge on at least two more occasions in the future.

Fern hunting is not without its dangers and the minutes of the 34th AGM held at the Ingleborough Hotel, Yorkshire on the 2nd September 1929 reported an accident which had befallen Dr F. W. Stansfield on the previous Saturday while fern hunting on the slopes of Ingleborough. He had slipped while scaling a wall and a large, sharp-edged stone fell upon him breaking or dislocating a rib. He was able to walk back to the hotel and he was removed to a nursing home at Lancaster the next day. It is reported that his absence and its cause had cast a gloom on the members but during the Monday they were cheered to receive a message from Dr Stansfield to say he was comfortable and there appeared to be no complications.

During the next decade the Society reached a plateau with membership at over one hundred in 1930 but reducing to 96 in August 1936. Several founding fathers were still alive such as R. Whiteside, T. Bolton, C. Henwood and W. F. Askew, but this small band continued to dwindle and a great loss was felt with the death of the great and much beloved Dr F. W. Stansfield on the 23rd February 1937. He held a unique place in the hearts and minds of the members. The effect of the good Doctor's death on the Society is reflected in the minutes of all the meetings recorded during 1937.

A Committee Meeting was convened by the President W. B. Cranfield at his office in Cheapside, London on the 23rd April 1937 'to consider what measures should be taken in view of the death of the Society's devoted Editor and Secretary F. W. Stansfield'. It was at this meeting that the decision was taken to strike a medal in memory of F. W. Stansfield 'to be awarded, on the directions of the Committee, to persons contributing to the advancement of the fern cult'. At the 42nd AGM held at the Coach and Horses Hotel, Charmouth, Dorset on the 19th July 1937, the President said that 'the Society had passed through a period of great anxiety as a result of the death of F. W. S. The Doctor was for many years Secretary and Editor of The British Fern Gazette; and it

Gleanings from the Minute Book 99

was owing to his unbounded industry and unlimited enthusiasm that the Society had flourished and gained in strength until it reached the position it held at the time of his death'.

However, although the ripples of the loss of F. W. Stansfield continued to affect Society business during 1938, this decade had also seen the emergence of new names that were to become the bones of the Society in the immediate post Second World War years. In 1930 A. H. G. Alston of the British Museum (Natural History) joined the Society and introduced the Society to this great institution which would provide such enormous benefits and set the Society on an ever broadening course of involvement with fern allies and contribution to the scientific study of world ferns. Others joining the Society were Percy Greenfield (1920), Reginald Kaye (1929), J. W. Dyce (1935) and Irene Manton (1936).

The 44th AGM was held at the George Hotel, Chard on the 24th July 1939 and once again W. B. Cranfield was re-elected President. He had now been President for 18 years and even he now admitted that the Committee should consider a younger man for the Presidency in the near future. Membership stood at 101 and the balance sheet showed £102 17s 3d (£102.83) in the funds. After the Society's business was over the date of the next meeting was set for the 22nd July 1940 to be held somewhere in the north of England, but not the Lake District, but that meeting was never held. Events in Europe were once again to overtake the Society and the Minute Book gathered dust for the duration of hostilities.

On the 16th September 1947, the President, W. B. Cranfield called a meeting at his premises at 6 Poultry, Cheapside, London. Present were Professor Weiss, A. H. G. Alston, J. W. Dyce, Rev. E. A. Elliot and P. Greenfield, the Honorary Secretary. I can do no better than to quote the minutes thus: *'A circular letter had been sent to members of the Committee by the Secretary in which the President described certain difficulties with which the Society was faced particularly as a result of the loss during the war of seven of the Officers, namely, Mr Blow, Alex Cowan, the Rev. Canon Kingshill Moore, Mr Sheldon (Vice-president), Dr T. Stansfield (Treasurer), Mr J. Sinclair (Auditor), Mr Lloyd and Mr Thorrington.*

'In the circular letter it was suggested that whatever other matters might have to be discussed, it seemed necessary (a) to appoint a Treasurer ad interim, *J. W. Dyce, who was elected to the Committee in July 1939 and was a bank official who was mentioned as willing to serve, and (b) to issue a further number of* Fern Gazettes, *due to subscribers, in order to complete the record of the Society's business up to the suspension of its activities in October 1939.*

'The members of the Committee unable to attend had either written to approve action as indicated under (a) and (b) or had raised no objection to it. It was accordingly decided to appoint Mr Dyce as Treasurer and to issue the British Fern Gazette *as soon as possible.'*

There was some discussion about the difficulties which would have to be overcome if the Society were to continue efficiently, but the Committee did not consider that it was within its powers to come to any major decisions which in its view must be left to a General Meeting. Information about the General Meeting and perhaps some forecast of the matters it might have to discuss would be given in the *Gazette.*

This was followed by a Committee Meeting on the 19th January 1948 at the same address, at which Reg Kaye was elected to the Committee and a long list of candidates proposed as new members, the majority of whom had been *'secured'* by the President. These were unanimously elected. The Committee was obviously keen to get the Society's affairs back to normal as soon as possible and it was agreed to publish a longer than usual *British Fern Gazette.*

Just when the Committee had thought it safe to poke their heads above the battlements, a letter was read to the Committee from A. J. MacSelf once again suggesting the name of the Society be changed to the British Fern Society! The minutes record that there was some difference of opinion on this question, which, in any case, would have to be raised at the next AGM.

A few months later the Society was to receive another blow. The President, W. B. Cranfield died on 29th May 1948. He had held the Presidency for eighteen years. Some felt this to be far too long as it had become the habit of the Committee for him to be re-elected 'on the nod'.

On Tuesday 10th August 1948 the 45th AGM was held in the restaurant of the Royal Horticultural Society's Old Hall, London. Robert Bolton was elected President with J. R. Pulham as Secretary and J. W. Dyce as Treasurer. The question of a change of name was deferred. Sadly, within a year Robert Bolton died and once again the Society found itself without a President.

In July 1949, A. H. G. Alston was elected President. One interesting item raised was the matter of W. B. Cranfield's fern slides; the matter was put in the hands of A. J. MacSelf with the suggestion that *'the Society keeps what he considered the best and the remainder to go to the Royal Horticultural Society for them to keep or dispose of as they thought fit'*. The last item of this meeting was the matter of A. J. MacSelf's suggestion of the change of name of the Society. It was agreed to bring this matter up at the next AGM.

The 47th AGM was held *'in the hut at the Southport Flower Show'* on 24th August 1950 with eleven members present. A. J. MacSelf's motion was discussed. *'After a full discussion those present were asked to vote. Ten were in favour of the motion which was accordingly declared carried.'* Yet in a brief note at the bottom of the minutes, *'This motion was withdrawn later by the proposer A. J. MacSelf and the seconder Prof. Weiss and the Society's name remains unchanged'*. Subsequent minutes fail to give an explanation as to why a motion so convincingly carried was withdrawn.

The Southport Show was the one item to feature prominently in the Society's business each year, and inserted into the minutes is a newspaper cutting from the *'Southport Visitor'* reporting the 1950 show. Under the headline *'Birkdale Man Gets Fern Trophy'* is a report on the 74 year old John Brookfield arranging a display of plants *'the public does not want these days'*. It goes on: *'Mr Brookfield was not interested in the taste of the public, he was in a fighting mood! He was after the British Pteridological Society's Trophy for the best group of hardy British ferns'*. John Brookfield is reported as saying *'the public do not want ferns these days, they do not want the trouble of looking after them.'* *'Also winner of the Sangster Trophy for the best display of ferns at Liverpool Flower Show, Mr Brookfield has more than 100 varieties of ferns – no two alike – in his display at Southport. . . . One of his most treasured ferns in his collection is* Athyrium filix-femina *'Clarissima', which he described as one of the finest ferns in the country, in fact it is so scarce that you cannot buy, borrow or steal one! I am only just getting an offshoot from it after 15 years.'*

The 49th AGM was held in the Society's tent at the Southport Flower Show on the 27th August 1952 with seven members present. No Committee Meeting had been held that year. The minutes recorded the death of A. J. MacSelf, *'sadly before he saw the publication of his book* British Ferns'. T. H. Bolton was congratulated *'on being the first member of our Society to broadcast a talk on fern growing: all of us who heard it will assure him of our enjoyment'*. This first broadcast and several articles on ferns published by members in the weekly garden papers, began to give the Society a much higher profile.

The AGM and Committee Meetings down the decade were poorly attended. At the 1954 Committee Meeting and subsequent AGM held at the British Museum (Natural History) only seven Committee members were present and the lack of a national programme was blamed on the fact *'that accommodation was expensive and difficult to secure'*. However, in 1956, the Society began once again to venture amongst British ferns. Two meetings were reported that year, a days's meeting at the Royal Horticultural

Society's Garden at Wisley and the first official post-war annual excursion which was to Kendal and supported by five members. This was a taste of things to come, for the minutes record that both meetings were recorded in the *British Fern Gazette,* the Kendal excursion '*most adequately by Mr Dyce*'.

So the decade of the fifties which at the beginning had seen a Society tentatively struggling to re-find its feet, came to a close with a Society brimming with new confidence. At the 56th AGM held in the boardroom of the British Museum (Natural History), London on the 26th September 1959, J. W. Dyce was appointed Secretary and re-elected Treasurer. The ailing Rev. E. A. Elliot had become a Vice-president. A. C. Jermy had taken over the editorship of the *British Fern Gazette.* A year later Mr Dyce's report began '*Our last AGM inaugurated a new era in the history of the Society, and we can look back on a year of progress, but marred by one sad event in the death of the Rev. E. A. Elliot after several years of ill-health*'. He went on to describe how the Society's scope had widened to include ferns and fern allies of the world and how closer links with the American Fern Society might lead to a larger overseas membership. The botanical expertise of the Society had been vastly increased by the pioneering work done by the new Editor of the *British Fern Gazette*, A. C. Jermy. So the die was cast for a modern Society prepared to embrace both the botanical and horticultural aspects of British and foreign ferns.

In 1962 the Society learned of the death of Robert Whiteside in November 1960 at the age of 94. So passed away the last of the founders of the Society. At the time of his death, he had been a member for 69 years which must stand as a record that will not often be broken. The Society still possesses his 1892 membership card.

By 1963 the membership had climbed steadily to 150 and the Society had settled into a busy routine of field and indoor meetings with regular venues such as Kew Gardens and the British Museum (Natural History). Northern meetings were organised by Norman Robinson who held the post of Northern Secretary. The practise of nominating and seconding new members had now been dropped and new members were elected *en bloc* at Annual General Meetings. The Society averaged 30 new members a year. As always there were resignations from the Society but a nett increase was still managed.

1963-4 saw the start of the Reading Circle which circulates the *American Fern Journal* to members.

In 1966 the Society organised a very successful fern foray to Austria, the Society's first official venture to the Continent.

In 1965 the Secretary, J. W. Dyce, introduced a single sheet newsletter to carry small items that either had not or could not be included in the *British Fern Gazette.* This had become a very erudite botanical publication with only the occasional horticultural item, which change caused contention amongst the amateur growers. The controversy came to a head in 1967. It was decided that the single sheet newsletter should be developed into a much larger publication, enabling it to be more comprehensive and leaving more serious matters to the *British Fern Gazette.* It was during this decade that the Society's meetings programme was developed in its present form. A Meetings Sub-committee was formed on which R. F. Cartwright and P. Temple volunteered to serve, and although the series of meetings was quite successful, the Committee did discuss why attendances fluctuated.

By the 66th AGM held at the Eagle House Hotel, Launceston, Devon on the 3rd September 1969, J. W. Dyce was pleased to report a busy and active Society numbering 327. The importance placed by the Society on an improved and vigorous programme of field meetings was indicated by the appointment of a Meetings Secretary as an officer of the Society, the post being held by R. F. Cartwright.

This period from 1959 to 1969 was the time when our Society met the challenge due in no small measure to the total commitment of J. W. Dyce as Secretary and Treasurer, and the Officers and Committee during those years. And so the Society entered its eighth decade with far more confidence than it had ever had. 1970 saw the introduction of a Spore and Plant Exchange scheme organised by David Russell, which was shortly taken over by R. F. Cartwright. A sad event in the same year, was the death of the Society's grand old man, Percy Greenfield. During 1973, the officers of the Society met Percy Greenfield's daughter-in-law (his son had died before him), Mrs Elizabeth Greenfield, to discuss how the Society should spend the gift of £500 donated by her in memory of her father-in-law, and so the Greenfield Fund was inaugurated to provide funds to help amateur projects in ferns and fern growing.

At the Committee Meeting held at the Royal Botanic Gardens, Kew on the 19th May 1973, the Committee were asked to consider a change of name for the Society. Once again, there was some objection expressed but after much discussion it was agreed to put this matter as a proposal to the membership at the next AGM. Immediately after this Committee Meeting, an Extraordinary General Meeting was called to consider and approve the Committee's proposal to raise the subscription from £1.25 to £2.

The 70th AGM was held at Capel Manor Horticultural Centre at Waltham Cross, Hertfordshire. After further discussion it was decided that the proposal for a change of name to the British Fern Society should be put to the membership in a postal vote. It was reported at the following AGM that 82 members had voted to keep the name and 55 had voted in favour of the change. The Society therefore retained its old name. A surprisingly small poll considering the membership in 1974 stood at 457. Perhaps now the Society could put this matter behind itself but events were to prove otherwise.

The early seventies saw the Society with an ever broadening membership and the need for more frequent postings to the membership became evident. By 1970 the newsletter had grown to include several pages listing reports of meetings, future meetings programme and news of general interest. J. W. Dyce also included articles of interest to fern growers mainly written by himself or occasionally a paper from a member. This proved very popular, so much so that the newsletter became the *Bulletin* in 1973. In 1975, due largely to the international flavour of the *British Fern Gazette*, the word 'British' was dropped from the title.

The early seventies saw a time of change for the Society. The annual general meetings had always been held in late summer, August or September, however, over a period of years, the AGM had become part of the autumn meeting in early October. This had led to problems in gathering in subscriptions, so it was decided to have a short year in 1975 so that the Society's financial year began on the 1st January with an AGM held during February 1976 although no officers or committee were elected at that particular meeting.

Throughout these years J. W. Dyce proved to be the fountainhead of the Society serving simultaneously as Secretary, Treasurer, Membership Secretary, distributor of journals and manager of book sales as well as attending every meeting each year. During 1975 Jimmy's health suffered and he realised that he must relinquish some of the burden. The Society was fortunate to obtain the services Dr B. A. Thomas as Honorary Treasurer, and later in 1976, those of Lt. Col. P. G. Coke as Membership Secretary.

In 1979 J. W. Dyce was elected President and I was asked to take over as Secretary, which I did with great reluctance, feeling that it was impossible to match Jimmy's unselfish and dedicated service. However it gave me the chance to meet new members and play an active part in the affairs of the Society.

It was felt that with the new decade of the eighties heralding the approach of the Centenary, the Society should carefully re-examine its role. The Minute Book records that at the Committee meeting held in May 1982 an item entitled *The Public Face of this Society* was considered, which was a discussion paper prepared by Nigel Hall. The paper argued, and the Committee generally agreed that the Society was '*failing its amateur growers*' and they went on to consider the possibility of introducing a new journal that would be predominantly horticultural. These considerations led directly to the introduction of a new journal, the *Pteridologist*, under the very able editorship of Martin Rickard in 1984.

All this talk of change inevitably led to the perennial question of the Society's name. After a lengthy discussion, it was decided to put the matter to a vote which proved inconclusive. Changing the name was again raised at the February 1983 Committee Meeting held at Aston University. After much heated discussion, it was finally agreed that the question should be put to that afternoon's Annual General Meeting seeking permission for the Committee to put the matter before all the membership with a postal ballot. This was agreed and the result of the ballot was announced at the AGM held in February 1984. The result was a convincing 'No' to changing the name of the Society: 98 for, 220 against and two spoilt papers. Has this result finally laid the matter to rest? It would seem so for the time being at least, but I would not like to put my money on it for the future.

In 1984 Richard Lamb revitalised the Plant Exchange Scheme as a separate entity and it is now run very enthusiastically by Rosemary Hibbs. The Spore Exchange had been built up into a very successful enterprise by Richard Cartwright with many members participating. In 1987, Richard Cartwright retired after 14 years of service and the Society was fortunate to obtain the services of Margaret Nimmo-Smith to carry on the good work. It is now the largest selection of fern spores available in the world. Conservation, both of species and varieties of ferns and fern allies has always been part of the Society's aims and has often been reflected throughout the Society's minutes. The 1980s saw increasing co-operation for the Society with the newly founded National Council for the Conservation of Plants and Gardens (NCCPG) and the establishment of the first National Collections of fern genera. Similar involvement is anticipated with Plantlife, founded in 1989 to protect endangered British habitats.

1985 saw the Society celebrating J. W. Dyce's 80th birthday with a presentation to him at a sherry reception during the afternoon of the annual Kew meeting, the day culminating with a superb celebration dinner in the Kew Gardens restaurant.

The minutes of the Committee Meeting held at Aston University in February 1984 were the last to be recorded in this venerable tome: it was a matter of a great deal of pride that it should fall to me to close this book after 93 years. I asked the Committee to consider the purchase of a similar minute book but they advised against it. So the minutes now reside in a modern lever arch file. It is disappointing that we shall pass on to posterity a nondescript file of type-written text: future generations will be denied the joy of ambling through such musty pages with joy and affection, assuming that a modern lever arch file will survive 93 or more years. However, on reflection, it is not the cover that matters but the stories contained therein.

Acknowledgements: I am most grateful to Margaret and Ian Nimmo-Smith for their assistance with the preparation of this paper.

'Matt' Busby is a skilled horticulturalist whose obsession with pteridophytes began when he was given the task of building up a comprehensive collection of ferns at Aston University for teaching purposes. His special interest is the Osmundaceae and he has an interesting collection of species and varieties. He has maintained the Society's presence at the Southport Flower Show for many years and, as Secretary of the Society since 1979, extends a warm welcome to new members.

The Plates

The following plates record all the British Pteridological Society's Presidents to date and show some of the Society's activities, officers and members. The railway plaque is included as a memento of the 'Victorian fern craze'.
Copies of all these photographs are held in the Society's archives and in the Botany Library, The Natural History Museum, London.

Josephine M. Camus

PTERIS TRICOLOR

F. W. Stansfield   C. T. Druery

P. A. Frazer   G. Whitwell   W. H. Phillips

W. Boyd   W. Gordon   W. B. Cranfield
J. J. Smithies   R. Whiteside   J. Edwards   W. B. Barker
H. Stansfield   G. Whitwell   T. G. H. Eley   C. T. Druery   G. Relton   A. Cowan
Arnside, 1902

W. B. Cranfield

R. Bolton

A. H. G. Alston

T. H. Bolton

photo courtesy of R.B.G., Kew

R. E. Holttum

R. Kaye

photo R. Kaye

J. A. Crabbe   J. Davidson

photo courtesy of University of Leeds

I. Manton

S. Walker

H. L. Schollick

J. W. Dyce

A. C. Jermy

photo A.R. Busby

G. Tonge

B. A. Thomas

photo A. Stirling

Old railway plaque near Ingleton, Yorkshire

photo courtesy of R.J. Smith

R. J. Smith   A. R. Busby   J. H. Bouckley
at the Society's stand, Southport Flower Show, 1988

P. Greenfield   T. Stansfield   T. B. Blow   W. B. Cranfield   W. Wilson   J. A. Dixon   R. Kaye   C. W. Smith
R. Whiteside
1938

photo courtesy of J.D. Osbourne & J. Ironside

R. P. H. Lamb A. C. Jermy P. J. Acock J. H. Bouckley A. C. Pigott A. GreatRex S. J. Munyard R. N. Timm
A. M. Paul   A. Hoare   M. Harris   J. M. Ide   M. Ladell
G. K. Hoare   P. M. Belsham
Birk Bank, North Yorkshire Moors, 1988

J. D. Lovis,　?,　C. R. Fraser-Jenkins,　T. Reichstein,　M. Gibby,　F. M. Jarrett,
I. Manton,　A. C. Jermy,　A. Sleep,　G. Vida
The Leeds Fern Group, Cambridge, 1974

photo M. Ladell

J. W. Dyce　　R. Kaye　　R. E. Holttum
at the celebration of Dyce's 80th birthday at Kew Gardens, 1985

The Society's stand in the scientific section at the Royal Horticultural Society's Chelsea Flower Show, 1991. The exhibit was awarded a Silver Medal.

photo P. Bloomfield

The participants of the international symposium on *The Biology of Pteridophytes*, Edinburgh, 1983

British Pteridology 1891-1991 117

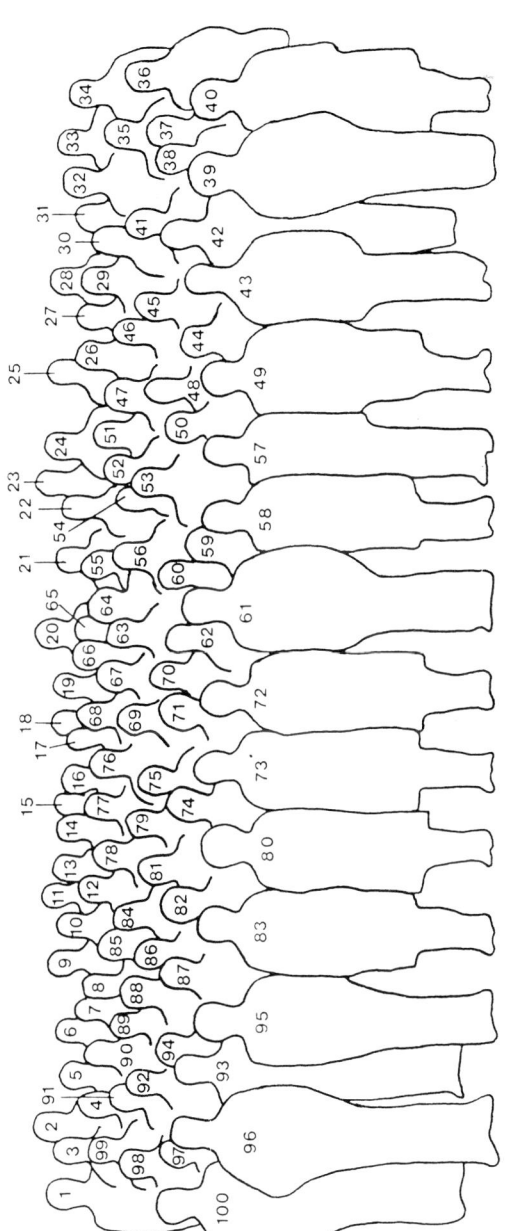

Key to the participants of the international symposium on *The Biology of Pteridophytes*, Edinburgh, 1983

Key to the participants of the international symposium on
The Biology of Pteridophytes, Edinburgh, 1983

1. C.N. Page
2. A.F. Dyer
3. R.L. Petersen
4. P.R. Hadfield
5. H. Schraudolf
6. Annette Fechner
7. G.M. Felippe
8. J.G. Duckett
9. D.M. Henderson
10. P. von Aderkas
11. A.R. Gemmrich
12. J.A. Raven
13. W. Greuter
14. D.L. Smith
15. L.G. Hickok
16. A.E. DeMaggio
17. J.E. Young
18. R.A. White
19. J. Schneller
20. A.C. Gibson
21. R.J. Johns
22. Angelique Hennipman
23. J.N.B. Milton
24. E. Hennipman
25. E. Klekowski, Jr.
26. B.G. Cumming
27. Kathryn Kavanagh
28. J.W. Merryweather
29. W.G. Chaloner
30. N. Quansah
31. D.R. Farrar
32. M.M. Yeoman
33. D.P. Whittier
34. D.G. Dunham
35. M.I. Cousens
36. Janet Dyer
37. Yvonne Herd
38. Alison Skene
39. N.L. Farrington
40. Marian Barker
41. J.T. Mickel
42. Elizabeth Sheffield
43. Karen Vondy
44. Alice F. Tryon
45. J.D. Montgomery
46. Susan Paterson
47. R. Viane
48. L.D. Gomez P.
49. Elizabeth G. Cutter
50. R.M. Tryon
51. J. Kornas
52. W.C. Taylor
53. D.S. Barrington
54. Brigitte Zimmer
55. P.R. Bell
56. C.H. Haufler
57. Judy Jernstedt
58. Joanne M. Sharp
59. I. Watanabe
60. Gerda van Uffelen
61. W.H. Wagner, Jr.
62. Diane B. Stein
63. S.M. Attree
64. T.G. Walker
65. Anne Sleep
66. W.J. Cody
67. P. van Hovenkamp
68. B. Øllgaard
69. J. Jubrael
70. Mary Gibby
71. Kate Tuckey
72. Florence S. Wagner
73. Jane Barker
74. Barbara M. Parkinson
75. V. Raghavan
76. T.A. Lumpkin
77. G.A. Peters
78. E. Wollenweber
79. H.E. Calvert
80. Reiko Yoroi
81. D.A. Stetler
82. Alix R. Bassel
83. Lesley R. Towill
84. C.F. Puttock
85. A.J. Willmot
86. J.H. Miller
87. Barbara Parris
88. A.C. Jermy
89. P.S. Pandey
90. M. Wada
91. R.B. Srivastava
92. A.F. Braithwaite
93. K. Mitui
94. Maina Padhya
95. M. Furuya
96. J.L. Ubera
97. Pilar Velez
98. S.C. Verma
99. R. Prange
100. A.P. Bennel

The Presidents of the British Pteridological Society
Nigel Hall
15 Mostyn Road, Hazel Grove, Stockport, Cheshire SK7 5HL

No society such as the British Pteridological Society could survive without the enthusiasm and dedication of a relatively small number of people who conduct the everyday business of the Society. 'Everyday' is, perhaps, not an adequate word to describe the range of functions performed, always voluntarily, by these officers. Across the Society's history these people have created policy, generated ideas, reconstructed, maintained viability in times of acute distress, offered advice, recruited, forged national and international contacts, and why – because of a commitment to things ferny. The passion, perhaps obsessiveness, that keeps these stalwarts going, has kept the Society alive. In some cases, as we will see, the devotion to the Society has meant taking on a quite extraordinary amount of work and it is something of a miracle that someone always emerges in times of need who is prepared to devote maximum time, and maximum energy, to ensuring that the Society has a future as well as a past.

In this brief overview I shall attempt to present short portraits of many of these people who have given so much of themselves to the Society. I shall concentrate (with two exceptions) on the Society's Presidents, but of course many others have also had major roles in the administration of the Society. There are clearly some people whose record demands greater space than that of others. I know that none of those 'others' will begrudge this amount of space; selflessness has to be a characteristic of those who work for the Society. Plus, we all recognise that the Society owes very special debts to certain people.

The Society traces its origins to a meeting on the 23rd September 1891. At that time the society was called the Northern British Pteridological Society. The original members elected a Chairman rather than a President. The next year a President replaced the Chairman.

J. A. Wilson (Chairman 1891-1892)

John Wilson (1831-1914) was one of the early Lakeland fern hunters, and his wife was also a fern hunter of repute. He was also a nurseryman and was noted by Druery as 'One of those striking characters whom to know was a pleasure and whose loss is irreplaceable'.

Dr F. W. Stansfield (President 1892-1897; 1902-1904; 1907-1908)

Frederick Stansfield (1854-1937) was the Society's first President. He was invited to take up the post by the founders. His connections with the Society were substantial and lengthy. He served (uniquely) as President on three separate occasions, as well as serving as Secretary from 1926-37 and Treasurer for a while in 1926, and Editor of the *British Fern Gazette* (1917-37). Only one other member in the history of the Society comes close to achieving this level of involvement in the Society's business.

The 'Doctor' was already part of a fern dynasty. His grandfather Abraham (1802-1880) had been the first member of the family to interest himself in ferns. From humble beginnings as a handloom weaver in Todmorden he started his own nursery and before long was stocking enormous quantities, and diversities, of ferns. Abraham clearly had an immense amount of energy, a quality which must have been inherited by his grandson, who as a child grew up in the Todmorden fern nursery. Frederick's apprenticeship in 'ferning' was substantial, and consequently his knowledge extensive.

Abraham also started the Todmorden Botanical Society and established in the family a respect for education. Abraham had several children two of whom, Abraham Jr. and Thomas, continued the interest in ferns. Thomas died soon after his father and his branch nursery in Pontefract was taken over by his sons, Frederick and Herbert. Frederick moved to Manchester to train as a doctor, and at the same time started a fern nursery in Sale. Once Frederick qualified in 1889 he moved to Derby and then to Reading. His brother Herbert then vacated the Pontefract nursery and took over the Sale nursery, which he ran until his death in 1928.

Frederick's professional concerns, which were intense and successful (he became Public Vaccinator for Reading, was President of the Association of Public Vaccinators of England, and President of a number of other medical societies), clearly did not inhibit his interest in ferns. His reputation and status in 1891 is unclear at this distance in time, but it was clearly considerable enough to be recognised by those pioneers in the Lake District.

During his editorship of the *Fern Gazette* he had the misfortune to have to record the deaths of several of his fern loving family; his nephew Walter in 1920, his wife Jane in 1926, his brother and fern nurseryman, Herbert in 1928, his daughter Elsie, wife of Percy Greenfield, in 1929, and his sister-in-law Sarah, in 1929. While F. W. Stansfield may not have had the journalistic talents of Druery, the first Editor of the *Gazette*, he was a much more elegant writer.

All the Doctor's contributions to the Society were a consequence of his extraordinary knowledge of ferns, and in particular, fern varieties. The Reverend Kingsmill Moore, in an entertaining reminiscence of the 'Doctor' captured something of his spirit when he wrote: *'The Doctor's devotion to ferns might almost be called a passion. He knew the origin of almost every extant variety, and where sources were obscure eagerly welcomed any clues. When a variety disappeared his regrets were as those for a lost relative'.*

In Stansfield's obituary, the writer, W. B. Cranfield, wrote that it was as *'a fern grower, hunter and raiser of choice varieties of our native British ferns that he will be remembered by our members'.* It is certainly the case that his involvement in these aspects of ferning was considerable, and indeed, through members who remember him, his influence continues. However, from a longer term perspective his work as Secretary and Editor deserves equal weight. To have taken on at the same time the two roles, even though he was retired, took an immense amount of time. To have continued doing this into his eighties is nothing short of astonishing. The Society remembers the 'Doctor' through the award of the Stansfield Medal.

Frederick's death was not the end of the family connection with ferns. His son Tom continued as a member and a committee member for a while (even contributing an article on 'ferns from Flanders' during the first world war) and his son-in-law Percy Greenfield maintained a lengthy connection with the Society, acting as Secretary from 1937-47, a critical time in the Society's history.

C. T. Druery (President 1898-1901)

Charles Druery (1843-1917), like Stansfield, deserves mention not simply because he served a period of time as President, but because it was his enthusiasm, and talents as a journalist, that led the Society to start a journal in 1909. Druery had been an acquaintance of Stansfield since 1882. He was described by Stansfield as a *'brilliant conversationalist, with an unfailing fund of wit and humour'.* Throughout his association with the Society he *'manifested an unflagging and whole-souled enthusiasm for the department of botany, which he had chosen as his speciality'.* He was known as *'Druery the fern man'.* He became a Fellow of the Linnean Society in 1885, and was awarded the Veitch Memorial Medal by the Royal Horticultural Society in 1897.

His testimonial is not just the volumes of the *British Fern Gazette*, but also three books: *Choice British Ferns* (1888), *British Ferns and their Varieties* (1910) and *The Book of British Ferns* (1903). The second book made available a large selection of miniature reproductions of the famous Jones' nature prints. The third book was the first British Pteridological Society book. The Society felt the need to produce a list of ferns, *'embracing only the pick and rejecting all such faulty ones'*. The Committee asked Druery to undertake the compilation and editorship of the book.

The *British Fern Gazette* came into being as a result of a suggestion from Druery, who, with a true recognition that the aims of the Society were not being realised, suggested that *'it was essential to arrange for greater publicity'*. Thank goodness the Society did agree with Druery, as our knowledge of the Society and its history would have been considerably poorer than it is. The Society had at that time about 110 members, so there are, inevitably, going to be very few complete sets of the Gazette around. Nevertheless the complete run represents a very special account of the Society's business and interests. It is somewhat inevitable in amateur societies that editors of journals end up doing a lot of the writing themselves, but Druery's journalistic skills were equal to the task. Druery's editorship lasted from the *Gazette*'s inception to his death on August 8th 1917.

Apart from his books and his editorship of the journal, Druery was responsible for the production of a number of papers for the Society between 1894 and 1905, authoring eleven of them himself. Like Stansfield, Druery's contribution to the formal business of the Society was a manifestation of his exceptional passion for ferns, fern growing and fern hunting. As a horticulturalist his main claim to fame was the documentation of apospory in ferns. He had, according to Stansfield, more success as a grower than as a hunter, and was instrumental in raising many beautiful and curious forms of ferns, although it is doubtful whether many of those remain in cultivation today. Stansfield and Druery can be said, both literally and metaphorically to have carried the Society into the twentieth century. They brought organisation and publicity to the Society which enabled it to continue successfully for many years.

G. Whitwell (Secretary 1891-1909)

One of the founders of the Society, George Whitwell (1839-1924) served as Secretary for 18 years. Unlike many other members of the Society who were 'gentlemen' or 'professional men', he was an ordinary working man with little pretension to education. He was, as Stansfield wrote of him, *'a gentleman of the first quality - truthful, honourable and kindly'*. Whitwell's forte was the ferns of the Lake District, of which he was a prodigious hunter and grower.

W. H. Phillips (President 1904-1907)

William Phillips (1831-1923) was a keen hunter and grower of ferns. Although living much (maybe all) of his fern hunting life in Ireland, he nevertheless was, until past his eightieth birthday, a regular attender at Society meetings. In his fern hunting career he found several score of varieties. He authored four of the pamphlets published by the Society prior to the creation of the *British Fern Gazette*. He published *Ferns of Ulster* in 1887 (with R. L. Praeger) and was also President of the Belfast Naturalists' Field Club 1905-7.

J. J. Smithies (President 1908-1909)

I have been able to trace little relating to James Smithies (*c.*1850-1931). He was the second founder member of the Society to stand as President and was noted in someone else's obituary as having been one of a group of North country *'fern hunters and raisers'*, and another founder member, Robert Whiteside, described him as *'a very keen grower and collector from Manchester, who came to live at Kendal and established a very fine collection'*.

A. Cowan (President 1909-1920)

Little is known of Alex Cowan's (1863-1943) activities. He was a paper maker at Penicuik, and also a hill farmer. He served as President of the Botanical Society of Edinburgh 1937-39.

W. B. Cranfield (President 1920-1948)

William Cranfield (1859-1948) was President for an extraordinary length of time, although it must be pointed out that this covered a period of eight years (around the Second World War) when the activities of the Society virtually ceased. He was almost the last President to serve for more than three years. Apparently every year when the Presidency came up for election it was felt 'good manners' to ask him if he would like to continue. Cranfield always said 'yes' and no-one ever suggested that he stood down for someone else. After his death the Society started to think about restricting the number of years anyone could serve as President.

An auctioneer by occupation, Cranfield had a wider interest in plants than simply ferns. He had a particular interest in daffodils, paeonies and irises, and was an active and prominent member of the Royal Horticultural Society, being awarded the prestigious Victoria Medal of Honour in 1935. He was also a Fellow of the Linnean Society. He served the Society in a number of capacities. He had terms as Treasurer and Secretary. He became interested in ferns as a result of his acquaintance with Stansfield and Druery. He did contribute a number of articles to the *Gazette*, and the *Gazette* carried several articles describing visits to his garden.

R. Bolton (President 1948)

Robert Bolton (c.1869-1949) was another member who, like Stansfield, was part of a horticultural dynasty. His uncles Henry and Thomas (a founder member of the Society) were both plant enthusiasts but in particular Henry Bolton (1845-1939) was noted as one of the most notable of the early Northern fern hunters. Henry Bolton, although not a founder member, joined the Society very early in 1892. Robert Bolton was a preeminent grower and collector. When President W. B. Cranfield visited Bolton's gardens in 1947 he wrote afterwards: *'It has been my privilege to see many collections, but never have I seen such diversity or extent'*. Robert Bolton was invited to be President when he was eighty, and recently bereaved. He was not in good health but accepted willingly, and, at the time of his death, had even been planning to attend the Summer meeting. The family connection with ferns continued and his son also became a President of the Society. Robert Bolton was awarded the Victoria Medal by the Royal Horticultural Society in 1949 for his skill in raising sweet peas.

A. H. G. Alston (President 1949-1958)

From its inception the Society had been primarily a horticultural, and amateur, one. After the war, the number of botanists becoming involved with the society grew and one of the consequences was the appearance, for the first time, of a professional botanist as President. From then on there has been a fairly equal distribution of Presidents between the two groups. Hugh Alston (1902-1958) joined the Society in 1931 and served as Editor of the *Gazette* from 1937. He was a distinguished botanist at the British Museum (Natural History), London, whose travels took him all over the world collecting ferns.

The Presidents

P. Greenfield (Secretary 1937-1947)

Although Percy Greenfield (1880-1970) was never President he might accurately be described as the best President the Society never had. He was the link between the pre-war members with their associations with the old fern hunters, and the post-war members of the Society. If Cranfield had not continued interminably as President then Greenfield would probably have been elected and would have maintained the Society's vigour much more successfully than did Cranfield. By marriage Greenfield, an official in the General Post Office, became part of the Stansfield dynasty and just as Stansfield was commemorated by the Society with a medal so Percy Greenfield is remembered by the 'Greenfield fund' set up to facilitate small projects relating to ferns. The fact that his term as Secretary spanned the Second World War period was critical for the recovery of the Society's fortunes after the war. Without his energy, experience, knowledge of ferns, and links with the past, the Society may never have recovered from virtual oblivion.

T. H. Bolton (President 1958-1960)

Tom Bolton (1899-1972), son of Robert Bolton, was President for only two years. Although part of a fern family, his particular passion was sweet peas and it was in this respect that he made his horticultural reputation. His son Robert was, for a time a member of the society but has concentrated on maintaining the family's professional horticultural interests.

Professor R. E. Holttum (President 1960-1963)

Professor Eric Holttum (1895-1990) initially studied Natural Sciences and then palaeobotany at the University of Cambridge. He lived in South East Asia for 32 years – first as Assistant Director and then Director of the Botanic Garden at Singapore and later as Professor of Botany at the University of Malaya – where he began his study of ferns. His famous *Ferns of Malaya* was published in 1955. On his return to England in 1954, he continued his research at the Royal Botanic Gardens, Kew, where he was editor of the pteridophyte parts of *Flora Malesiana*, and published papers on many groups of ferns. Reknowned also for his work on orchids, given two honorary doctorates and recipient of many medals and awards from learned societies, Holttum ranks as the most erudite tropical fern taxonomist this century.

R. Kaye (President 1963-1966)

Reginald Kaye belongs firmly to the horticultural tradition of the Society. He remembers being interested and intrigued by ferns even when as young as six years old. In 1937 he took over a nursery and really began to acquire a comprehensive collection of British ferns and their varieties (see Kaye's article in this volume). He possesses in his private garden many of the ferns collected by the old Lakeland fern men. He is the author of the only major recent book to specifically deal accurately and comprehensively with the growing of hardy ferns. It offers a distillation of a lifetime's experience of growing ferns.

Dr J. Davidson (President 1966-1969)

James Davidson (d. 1985, aged 89), originally a forensic scientist, had a lasting passion for alpine plants before he became interested in ferns. He also served as President of the Scottish Rock Garden Society.

Professor I. Manton (President 1969-1972)

Professor Irene Manton (1904-1988) was a distinguished botanist. After studying at Cambridge she eventually became a lecturer at Manchester University where she started studying the cytology of ferns and after the war she published the seminal work *Problems of Cytology and Evolution in the Pteridophyta* (1950). She took up the Chair of Botany at Leeds University in 1946 where she studied the ultrastructure of plants with the recently invented electron microscope. Five universities conferred honorary doctorates on her. She became a Fellow of the Royal Society (1961) and President of the Linnean Society (1973-76). Manton joined the British Pteridological Society in 1936 and published articles and notes several times in the *Gazette*. Irene Manton was the Society's first woman President. It is noticeable that it is was only after the botanical side of the society increased did many women become members. Before then it was a fairly exclusive middle-class male club. Manton's other consuming passion was art and she bequeathed, through the Linnean Society, an annual prize of a piece of sculpture or fine art and a sum of money for the best botanical doctorate thesis.

H. L. Schollick (President 1972-1975)

Henry Schollick (1906-1991) was a publisher with Blackwells, and perhaps this professional interest in books contributed to his passion for books on ferns. He had probably the finest private collection of fern books in the country. He also had a large and select collection of ferns and was one of the few Presidents who was a regular supporter of the Society's stand at the Southport Flower Show.

Dr S. Walker (President 1975-1979)

Stanley Walker (1924-1985) was a student of Irene Manton's at Leeds University where he studied the cytology and taxonomy of *Dryopteris* ferns. He then became a lecturer in botany at Liverpool University. He later helped establish a department of genetics at Liverpool where his interest shifted from plant genetics to human genetics. (One of his ex-students, Mary Ambrose [now Dr Mary Gibby] has contributed an article to this volume.)

J. W. Dyce (President 1979-1982)

I have already mentioned the debt that the Society owed to Percy Greenfield for his role in revitalising the Society after the Second World War, but it is to Jimmy Dyce that the major thanks of the Society must go. One has to go back to Stansfield and Druery to find an equivalent dedication, degree of effort, knowledge about ferns, and indefatigable and total dedication to the Society.

Jimmy Dyce became a member of the Society in 1935 after becoming intrigued with ferns. When shortly afterwards the Society required an auditor, Jimmy, a banker, was able to step into what was to be the first of many administrative roles. After the war it was Jimmy Dyce whose letter to Percy Greenfield resulted in a meeting being called of remaining members. As Jimmy put it *'It was attended by six members . . . It was reported that seven of the Society's officers had died during the war, and the general feeling was that the Society was dead and should not be revived. Mine was the only dissenting voice'*. Fortunately (and as many of us know) Jimmy's dissenting voice contains powerful advocacy; the Society was revived.

Jimmy became Treasurer, which at that time included Membership Secretary. By the end of the year the Society had acquired 132 members. Membership revolved around the 100 mark for several years, and when the Secretary, the Rev. Elliot, died, Jimmy added to his chores by taking on the Secretary's role. At that time the editorship of the *Gazette* was taken over by Clive Jermy (see below). The Society began to grow in numbers, but the development as a botanical society upset some members who were of a less scientific mind. By then Jimmy had retired and with his extra time attempted to maintain the balance by enlarging the newsletter. He did this so successfully that six years later it became the *Bulletin*. At the same time as all this he was busy arranging meetings programmes and leading ferning expeditions to all corners of the country. In addition he started BPS Booksales and was mainly responsible for the distribution of the Society's two journals.

It seems quite incredible that any one person could reasonably have carried out all those burdensome duties. However, Jimmy did so, and always with meticulous accuracy, scrupulous honesty, and with the utmost friendliness. There are many people across the UK and the world who feel that Jimmy is a personal friend, so warm was the response when they inquired about membership or any other details about the Society.

By 1975 Jimmy began to divest himself of some of those burdens. Thanks to him the Society was large enough that there were competent and willing people able to take over the arduous jobs he had been doing. In the end it took about five people to handle all the work Jimmy had been doing on his own. Many people would at this point have sat back and taken things easy. Not so Jimmy, who with additional space and time continued and developed his correspondence, developed the Booksales, travelled to the USA, tended his ferns, wrote articles for the *Bulletin* and subsequently the *Pteridologist*, contributed *Fern Names and their Meanings* and *The Cultivation and Propagation of British Ferns* to the Society's 'Special Publication' series, and offered advice, help and friendship to all fern enthusiasts. He is still doing so and we hope will go on doing so for many years yet. He was elected the Society's first President Emeritus in 1985 and has been a Fellow of the Linnean Society for many years. He has contributed an article on the history of the Society to this volume.

A. C. Jermy (President 1982-1985)

Clive Jermy joined the Society after the war and was Alston's successor as Head of the Pteridophyte Section at the British Museum (Natural History). When Clive Jermy took over the editorship of the *Gazette* the direction of growth of the Society changed fairly dramatically. The *Gazette* became primarily an academic botanical publication, and the BPS began to be recognised as a scientific society. This change probably resulted in the biggest ever growth in membership of the Society as botanists across the world joined, and the Society really became fully international with its third international symposium to be held in 1991. His concern with worldwide conservation issues was instrumental in founding the British conservation organisation, *Plantlife*.

G. Tonge (President 1985-1988)

Gwladys Tonge was only the second woman President of the Society. She is a keen plantswoman who has also been Chairperson of the Hardy Plant Society, and a very active member of the British Ivy Society. She is an accomplished artist and has exhibited her work (and been awarded a Gold Medal) at RHS shows. She has published a gardening autobiography, *A Gardener's Progress*.

Dr B. A. Thomas (President 1988-1991)

Barry Thomas, who is currently Keeper of Botany at the National Museum of Wales, has had several roles in the Society, acting as Treasurer and Membership Secretary for 15 years. He was instrumental in establishing our third journal, the *Pteridologist*. His particular interest is in fossil pteridophytes and the evolution of plants with many papers and books published on this subject.

This brief summary of the Society's Presidents inevitably leaves out many people who held other posts within the Society. Many of these contributions are relatively recent and no doubt the Society's second centenary publication will record their achievements. Looking back over the list one thing is striking: Presidents of the British Pteridological Society seem to manage to live to a very ripe old age. There must be something about an interest in ferns which encourages longevity. Let us hope it continues to do so.

Nigel Hall enjoys growing ferns and has been an avid collector of fern books. His particular interest in the social history of ferns and fernists led to him being appointed the Society's first Archivist in 1980. Nigel is a lecturer in, and writes about, early years' education at Manchester Polytechnic. He is also the author of several children's books.

Testament of a Fern Lover
(Dedicated to all fern lovers, past present and future)

Whenever you see a host of ferns
In full array of crozier display
Think of me, occasionally
Not with sorrow,
Or great loss
But with affection and,
No fuss please, for I am fortunate
And have observed,
However well or un-deserved
The delicate tracery and infinite form
In spinney and combe
In sunshine and storm

Of one of earth's oldest miracles.

I have experienced the perfume
Of a leafy dell,
The mystery of a hidden cleft and
Walked sodden ground
Midst clinging thorn and darkened glen
Where they are found, and
Stood elated on rocky hill and ancient quarry
To observe, with pleasured eye
Their rare majesty.
I have searched for nature's gifts with pleasure
Found peace in quiet shady places and
Lifelong companionship and friends.

I am content.

R. J. Smith
184 Solihull Road, Shirley, Solihull B90 3LG

Ray Smith first became interested in ferns in the early 1960s when walking in Scotland. Now retired after a career in Personnel Management, he is a keen grower and respected exhibitor of ferns. He is interested in all species and varieties of ferns, but confesses to having a soft spot for Polypodium *and* Polystichum *varieties, with* Osmunda *and* Adiantum *close behind.*